Birds as Builders

By the same author

The Birds of Saltram
Projects with Birds

Birds as Builders

Peter Goodfellow

Arco Publishing Company, Inc. New York

To David and Andrew

Even the sparrow finds a home,
and the swallow has her nest,
where she rears her brood beside thy altars,
O Lord of Hosts, my King and my God.
Psalm 84 v.3

Published by
Arco Publishing Company, Inc.
219 Park Avenue South New York
N.Y. 10003

Library of Congress Cataloging in
Publication Data
Goodfellow, Peter Birds as Builders
1. Birds–Eggs and nests. I. Title.
QL675.G62 598.2'5'6 76–53802
ISBN 0–668–04183–8

Printed in Great Britain

Contents

Preface

In 1911 Francis Herrick wrote, 'In many ways it would be difficult indeed, from the standpoint of instinct and behaviour, to find a more unsatisfactory class of scientific literature than that which deals with the nests of birds.' Since then much progress has been made in the study of birds' nests, especially in areas other than the Holarctic, and although several general works on ornithology have summaries of this knowledge, most of the interesting details are scattered through a vast literature. This book attempts to describe the wide range of birds' nests, a subject which greatly intrigued the general reader of Natural History a century ago but which has seldom since been comprehensively described.

I am very conscious of my great indebtedness to all the birdwatchers whose observations fill the book. To them and the book's illustrators I extend my sincere thanks. My special thanks are due to Dr C. J. F. Coombs, who worked so hard to produce such excellent drawings to my requirements, to Roger Hosking, who went out of his way to take some of the photographs, to Graham Madge, who loaned me several precious nests from Malaya, and last but not least, to my wife, June, who was my constant source of encouragement and able typist. The book succeeds through them, and fails if I have not heeded Herrick well enough.

1 What is a nest?

As well as birds many insects, fish, and mammals build nests, so an ornithologist has to be careful how he defines the word 'nest'. Birds are the most expert and most industrious nest-builders, although it must be remembered that some birds build no nest. A comprehensive definition of a bird's nest is needed, such as: 'a structure or excavation made by birds, or the modification of a structure or excavation already in existence, or any place in which eggs are laid and are incubated until hatching'.

Spotted flycatchers *Muscicapa striata* **are well-known for their unusual nest-sites, often choosing man-made objects for the nest platform.** (*D. N. Dalton/NHPA*)

The function of the nest is obviously to protect the eggs and young from bad weather and predators. This is especially true of species rearing young which are helpless at birth (ie nidicolous chicks). Many ground nesting birds, such as terns, rear young which, on hatching, are already covered in down and which are mobile as soon as they are dry (ie nidifugous chicks). These birds have little or no nest; a hollow in the ground is sufficient to hold the clutch together, facilitated, in waders especially, by cone-shaped eggs; a clutch of four wader's eggs fit neatly together when the pointed ends all face inwards.

Wrens share this apparent preference for the unusual. Here one is about to leave its nest in the sleeve of a discarded jacket. (*William S. Paton*)

Table 1

Analysis of the means through which protection is secured by birds during nesting

1. Through instinct
   - (i) concealment
     - (a) through special instinct, such as soiling eggs with excreta by eiders.
     - (b) by natural cover as in a cave, hole, or dense foliage.
     - (c) by artificial cover, such as that used by grebes, or such as the outer, disguising fabric of a nest.
   - (ii) care of eggs or young.
   - (iii) defence of nest or young by fighting.
2. Through intelligence, modifying instincts.
3. Through structure
   - (i) the external form of eggs, young or adult with or without the aid of colour, eg many species' use of camouflage.
   - (ii) the physical modifications of adult or young as in the specially strong feet and claws of young fairy terns.

A bird's nesting is actually part of its life cycle which might be: migration, courtship, mating, selection of nest site, nest building, incubation, care of young, fledging, migration . . . It could be argued that to isolate nesting from the rest of behaviour is artificial. Although much work has been done studying birds to learn what nest-materials they use, to understand their stereotyped building behaviour and to discover what stimulates a bird to build, little or nothing has been done to relate all the different kinds of birds' nests to the behavioural patterns. Herrick (1911) stated that the classification of birds' nests should be made on the basis of behaviour. He proposed three classes – the third being the highest. He offered his classification for discussion and several points of interest may be noted. Because he believed that birds making no efforts to protect or conceal their eggs were not nest-builders, he did not

Peregrines are primitive nest builders, making a rough scrape and possibly adding a few wisps of straw. (*William S. Paton*)

include these species in his classification. He considered the megapodes made primitive nests, yet compared with the rest of their order, the *Galliformes*, some of their nests are extraordinarily complicated, although burying eggs, like crocodiles, *is* more primitive than incubating them. His third and highest class is much the largest and is subdivided 19 times; it is comprehensive but clumsy. Surprisingly, the detailed survey by Marchant (1964) makes no mention of it and his classification is based on subdivisions of the two groups, non-passerines and passerines. A classification, based on Herrick's scheme has much to commend it and a suggested adaptation is shown in Table 2.

Table 2

Classification of the nests of birds on the basis of behaviour

1. Primitive nests
   - (i) no nest at all, eg nightjars
   - (ii) a nest-scrape, perhaps with a few straws, eg peregrine
   - (iii) hole and mound nests of magapodes

2. Secondarily adaptive nests
   (i) nests in natural or artificial cavities, eg owls, parrots
   (ii) nests in partly adapted cavities, eg hornbills
   (iii) using other birds' nests, eg some raptors
   (iv) building simple platforms, eg pigeons
3. Primarily adaptive nests in which the greatest efforts for concealment and protection are made
   (i) excavated nests, in earth or wood, eg kingfishers
   (ii) constructed nests (a) 'standing type', eg thrushes
   (iii) constructed nests (b) 'hanging type', eg weavers
   (iv) communal nests, eg quaker parrots.

Many loopholes may be found in this; eg should 2 (iv) really be part of 3 (ii)? And should titmice be classified under 2 (ii) or 3 (ii)? This book attempts to be simple and easily comprehen-

A female blackbird *Turdus merula* with young. The female usually takes the initiative in nest-site selection. *(E. V. Breeze Jones)*

The kittiwake chooses to nest on cliff ledges building a substantial but simple platform. (*William S. Paton*)

sible, although it, too, fails to describe precisely some nests, such as warblers' *Phylloscopus spp*, which build enclosed nests (Chapter 5) on the ground (Chapter 8). Herrick from his original would have had no difficulty, however; they would have been primarily adaptive, independent, constructed, standing, wholly constructed; or in his shorthand (which he used in his text), 3, i, b, l, (a)[1]. Precise, but a headache to understand! The scheme of this book is based on nests' physical characteristics, which are probably the criteria most birdwatchers would use; and the sequence 'no nest – simple – cup-shaped – enclosed – hanging and woven' mirrors what is probably the evolution of the bird's nest (Makatsch 1950).

Before any bird has a nest, however primitive, the pair has to decide where to nest, an aspect of behaviour called 'nest-site selection'. In many species the male and female share the task, with the female taking the initiative more than the male, as is the case in the blackbird *Turdus merula* (Snow 1958), and red-necked phalarope *Phalaropus lobatus* (Tinbergen 1953) which then leaves the incubation and rearing of the young to the male. In contrast it is the cock house sparrow *Passer domesticus* who chooses the nest-hole, calling and displaying at its entrance, although final selection is made by the hen, who chooses a mate with a suitable home. A cock with an unsuitable site will find difficulty in pairing (Summers-Smith 1963). In the blue tit *Parus caeruleus*, the proposal to use a particular site is put forward by the male performing a series of ritualised displays. This may well begin in late December although the nest is not built until April (Howes 1969). Herring gulls *Larus argentatus* share the choosing, and the Scottish crossbill *Loxia pinicola* has no firm pattern, each pair having its own method of nest-site selection which possibly depends on the particular sexual and building rhythms of the two birds (Nethersole-Thompson 1975).

Nest-building behaviour is equally interesting, and is instinctive, although it has been suggested that instructive demonstration plays some part: the builder has to learn where suitable materials may be found, and it may even learn how to improve its building technique. Birds which build the simplest nests use a single movement, 'sideways-throwing', to accumulate material. In the modified form, 'sideways-building', material is arranged on the site. Birds stand up to build in both cases, but an incubating bird will often perform 'sideways-building'. Both types of construction are found in petrels, ducks and geese, pheasants, waders and gulls. Sideways-throwing often occurs by the nest-site at nest-relief, and is performed by the bird leaving the nest – a few pebbles, animal droppings or grass,

Above:
A rare photograph of a well camouflaged hazel hen *Tetrastes bonasia* **in her simple nest on the ground.**
(J. B. & S. Bottomley)

Right:
Lapwings' *Vanellus vanellus* **cryptically-coloured eggs make a neatly-fitting clutch.**
(E. V. Breeze Jones)

This white-fronted bee-eater *Merops bullockoides* belongs to the most skilled group of builders and makes its nest either by excavating or by adapting a hole in an earth bank. (*J. F. Reynolds*)

depending on the site. Sideways-throwing is a very ineffective form of nest-building. The first stage beyond it is carrying material to the nest-site (eg some waders and terns). Sideways-building is a considerable step forward in the evolution of nest-building. A rampart of material is put up around the incubating bird as she moves and turns on the eggs (eg swans and cranes). The material surrounds a hollow the size of the bird's body, which, although often simple, is immediately recognisable as a nest. Sometimes sideways-building movements produce concealing arches of grass over waders' nests. Very little modification of sideways-building would be needed to enable a bird to create a more effective form of nest-building (Harrison 1967). Among passerines, which build some of the most complicated nests, several movements are common to many species. Loose strands of material projecting from the

The little tern ***Sterna albifrons*** **nests in small colonies, usually just scraping a hole in the sand or shingle.**
(*E. V. Breeze Jones*)

rim are pulled into the cup by 'pulling and weaving'. The builder will push down the cup by 'scrabbling' with her feet, and by 'turning' while sitting in the nest she will shape the cup with her breast. Weavers have even more specialised actions (Chapter 6).

The role of the sexes in nest-building varies considerably, as the following list adapted from Ryves (1944) shows:

1. Both sexes build equally, eg woodpeckers and kingfishers.
2. Both sexes build, the male constructing the main fabric and the female doing the lining, eg wren.
3. The female builds with material provided by the male, eg wood pigeon.
4. The female collects material and builds, eg vireos and tits.
5. The female builds with material collected by both sexes, eg wall creeper and rook.
6. The male alone builds, eg weavers.
7. The male builds with material provided by the female, eg frigate birds.

Many birds have been encouraged by naturalists and bird-lovers to breed in nest-boxes. Many hole-nesting species take to them readily, providing delightful additions to the life in many suburban gardens, and also easily observed nests for ornithologists to study. Sometimes the latter have erected boxes in a woodland area and have been able to study in detail the breeding biology of birds such as pied flycatchers *Ficedula hypoleuca*, which would otherwise have been impossible.

Finally, the fascination of searching for and looking at birds' nests is never-ending because of the comical discoveries one often makes – nests made of strips of polythene instead of grass, a pigeon's nest of lengths of wire instead of twigs, a robin's nest in a thrown-away kettle, a spotted flycatcher's nest on the cistern of an outside toilet, a gull's nest in a fish-box above the high-tide line, a wren's nest in a skull, a blue tit's nest in the street lamp outside my house (it ignored a perfectly good nest box in the back garden!) and many, many more.

2 No nests

The most highly evolved species of birds, the vast group known as the passerines, generally build fairly substantial nests. Non-passerines make simple ones, or do without a nest at all, as did the birds of prehistoric times. Birds and their nests have come a long way since the days of the earliest known bird, *Archaeopteryx*, but several species from a variety of families still simply lay their eggs on the ground or in natural cavities.

The ostrich *Struthio camelus*, from the world's most primitive living order of birds, the Ratites, provides a good example of the simplest of 'nests'.

The male may have several hens, and they use a communal nest. When the site has been selected, the ostriches clear an area in the grass by trampling down the herbage, then scrape a hollow in the soil or sand, about 35½in (90cm) in diameter and 11¾in (30cm) deep. Each lays six to eight eggs here. As the

Kittlitz's sandplover buries its eggs in the ground to keep an even temperature. (*J. F. Reynolds*)

When the eggs are uncovered the incubating sandplover will shade the eggs from the sun. (*J. F. Reynolds*)

hollow does not give much protection the incubating bird may stretch its head and neck along the ground in the effort to become inconspicuous. Many eggs will nevertheless fall prey to prowling hyenas and jackals. Others may not hatch because the incubating bird cannot cover the clutch – which is often of 20 eggs, sometimes 50, and even 80 have been recorded.

An ostrich's egg is remarkable, despite the poverty of its nest. The largest laid by any living bird, it measures about 6in × 5in (15cm × 12½cm), weighs 3lb (1⅓kg) and has the volume of about 20 hens' eggs. Despite its size, it is in fact a small egg in proportion to the size of the adult female, being only 1½ per cent of her weight, whereas in a large number of other species the egg weight is between 7 and 11 per cent. Strangely, eggs are sometimes found away from the nest: it is believed that they are allowed to hatch in the heat of the sun. They are, perhaps, among the first in the clutch (which may take two weeks to be laid), and are set apart so that they do not hatch earlier than the rest. If this is so, it shows an amazing instinctive action by the adults to improve the chances of successful breeding – which on the face of it gets a poor start from the bare nest-scrape.

Above:
The nightjar *Caprimulgus europaeus* makes no nest at all but relies on its remarkable camouflage to protect it from predators while incubating.
(Roger Hosking)

Right:
The greater potoo *Nyctibius grandis*, which is related to the nightjar, also relies on camouflage. After laying its egg in a depression on the top of a vertical tree stump the incubating bird sits upright on top of it looking remarkably like a continuation of the wood.
(Dr. C. J. F. Coombs)

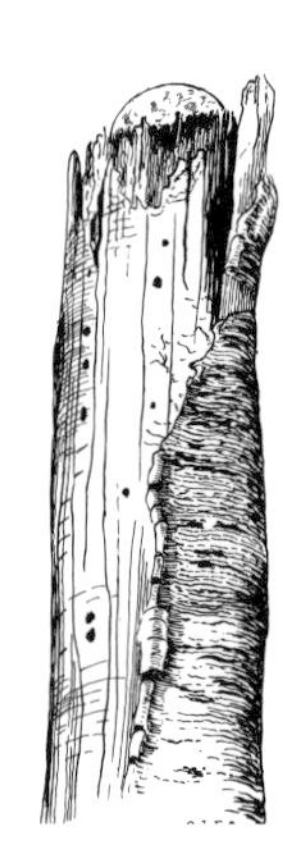

The nightjars *Caprimulgidae* do not even make a scrape; their nest is non-existent. They all lay two eggs, directly on the ground, usually in heather or bracken. Although that seems to give them no protection, the miracle is that the incubating bird is so well camouflaged that the eggs are in fact safe. Very often the European nightjar *Caprimulgus europaeus* chooses a site near a log or fallen branch: because the bird looks like wood, the real log adds to the disguise. The bare nest sites must have more merits in the birds' eyes than ours, for they have been recorded as being used several years running – the eggs being laid within a few centimetres of the previous clutch.

The common potoo *Nyctibius griseus*, from South America, is a member of a family closely related to nightjars, and its nesting habits are even more curious. It roosts in an upright position on a tree stump, looking so like part of the broken wood that it is easily passed by. Its single egg is laid in a depression on the top of a broken-off branch, or in a hollow on a sloping branch, fitting the depression so well that it is difficult to extract it.

Several owls neither make a nest nor use a hole in a tree. The short-eared owl *Asio flammeus* lays its eggs in a hollow, in trampled vegetation in open country. The young owls, when about two weeks old, explore the area and may wander up to 100 metres from home. The closely related long-eared owl *Asio*

The short-eared owl *Asio flammeus* likes to lay its eggs on trampled vegetation in a hollow in open country. (*W. S. Paton/Aquila*)

The wood sandpiper *Tringa glareola* sometimes nests in a tree, taking over, in this case, an old song thrush nest.
(J. B. & S. Bottomley)

otus, a bird of woods and forests, also makes no nest of its own, invariably using the old nest of a crow, jay, magpie or even wood pigeon; squirrel dreys will serve too.

The green sandpiper *Tringa ochropus* takes over an old thrush, crow or pigeon nest, adding little or no new material. It is curious that this wading bird (and sometimes its close relative the wood sandpiper *Tringa glareola*) should be a tree nester whereas practically all other waders are ground nesters.

Some waders do build rudimentary nests of grasses, but many lay their eggs on the bare ground. Species such as plovers prepare a scrape for the eggs but others make none at all. Occasionally one bird may ring the eggs with stones or animal droppings, but another of the same species may do without any decorations. The eggs of waders are wonderfully camou-

The fairy tern *Gygis alba* lays a large, single egg in a depression on a horizontal branch or on a cliff ledge. (*Dr. C. J. F. Coombs*)

flaged, their grassy, sandy or stony colours covered with darker freckles, often causing them to be almost invisible on the ground. There is an interesting relationship between egg-colour and ground-colour: eggs on pebbles or sand generally have a light base colour; eggs in grass, marsh or moorland vegetation have usually a much darker olive background with heavier markings. Kittlitz's plover *Charadrius pecuarius*, which is found in much of Africa south of the Sahara, often goes one step better by covering its two eggs with sand or grit when it leaves its nest. It does this with a very rapid shuffling movement of its feet (Hall 1960).

Terns mostly nest on the ground, but the fairy tern *Gygis alba* is a curious exception. The bird has entirely white plumage, save for a black ring round its black eye, and is therefore

strikingly visible against any background. Yet its breeding site is on a cliff or horizontal tree branch. No nest is made; the cryptically coloured egg is laid on the stone or wood and balances there. On Ascension Island, Dorward (1963) found that nearly always the egg was laid on a spot too small to take a second one. Indeed, the egg is very large and only one can be brooded.

The species has evolved two special adaptations: a cautious way of settling on and rising from the egg, and sharp claws on the chick's toes for gripping. Many eggs are successfully incubated on what appear to be very awkward places. Dorward watched an egg being laid precisely on the spot where it was to remain – it could not have been moved without falling. There was an audible click as it was deposited. Some eggs do of course crack as they are laid and fail to hatch. But the adult fairy terns are very attached to their eccentric nest site; an egg moved only a few centimetres may be ignored. Ringing has proved that pairs will return in following seasons to breed on exactly the same tiny ledge or the same notch on a tree-branch. Despite these precarious nurseries, the species is remarkably successful. Indeed on Ascension Island it was discovered that its breeding success was higher than that of most other species there.

A successful species with more bizarre nesting habits is the emperor penguin *Aptenodytes forsteri* of the Antarctic. Some

Above left:
The cream-coloured courser *Cursorius cursor* brooding in the heat of Africa, uses the techniques of opening its bill, ruffling its feathers, spreading its wings and exposing its legs in order to keep cool. (*J. F. Reynolds*)

Below left:
The courser frequently lays its eggs among animal droppings as the colour and pattern of the eggs make them especially inconspicuous there. (*J. F. Reynolds*)

The male emperor penguin *Aptenodytes forsteri* is the incubator of the pair. After the female has laid her single egg he will gather it up on top of his feet and keep it warm by covering it with the fold of skin on his belly. (*Dr. C. J. F. Coombs*)

penguins do make nests of a pile of stones, or lay their eggs in burrows, but when the emperor female has laid the single white egg on the ground, she returns to the sea to feed, and replenish her supplies of fat. The male takes charge of the egg by carrying it on top of his feet against his shanks, keeping it warm by covering it in a fold of skin on his belly. He must stand in temperatures down to —32°F (—35·6°C) in icy winds and blizzards, and by using this unique adaptation the emperors manage to breed in the Antarctic winter. Scientists have discovered how efficiently the male transfers heat from his body to the egg. When external temperatures were —9°F (—22·8°C), the internal egg temperature was 82°F (27·8°C) which shows the efficiency of the insulation of skin and feathers. He lives on his reserves of fat for about 60 days until the female returns to relieve him when the chick is due to hatch.

The curlew *Numenius arquata* is a ground nester, sometimes lining the nest-scrape with grass. (*Roger Hosking*)

Diagrams comparing visibility of a double-banded courser *Cursorius africanus* nesting in a hollow (a) and on a rise (b). In the hollow it can see over the shrubs, whereas on the rise its view is impeded. (*Maclean 1967*).

3 Simple nests

The birds which build very simple nests, one stick or reed or stone upon another, are all non-passerines. Although of rough construction some of these nests are large.

The Adélie penguin *Pygoscelis adeliae*, the most abundant and widely distributed of all Antarctic penguins (Sparks and Soper 1967), builds a nest of stones, piling them up until the cup is raised well clear of the ground. The male builds the mound; if a female accepts one of the last stones, she becomes his mate. The construction is a protection against flooding. Nests are built and reinforced in frantic attempts to keep chicks dry during midsummer storms, when snow turns to rain and

A male chinstrap penguin *Pygoscelis antarctica* makes a gesture of appeasement or greeting to his female by bringing a stone and dropping it in front of her. (*Niall Rankin/Eric Hosking*)

the ground becomes slushy and puddled. Adélies have a high regard for stones; besides using them for building, one sometimes brings a stone to its mate as a gesture of appeasement or greeting. As stones become eagerly sought in the vast penguin rookeries thefts are rife. Levick (1914) presented distinctively coloured pebbles to some birds, and soon these were much sought after and became distributed throughout the rookery. An early Antarctic explorer, Louis Bernacchi, in 1899 described a robbery thus:

> The thief slowly approaches the one he wishes to rob with a most creditable air of nonchalance and disinterestedness, and if, on getting closer, the other looks at him suspiciously, he will immediately gaze round almost childlike and bland, and appear to be admiring the scenery. The assumption of innocence is perfect; but no sooner does the other look in a different direction, than he will dart down on one of the pebbles of its nest and scamper away with it in his beak.

Perhaps the simplest tree nest of twigs is that built by many species of doves and pigeons, such as the wood pigeon *Columba palumbus*. The prime requirement for the site seems to be a platform of flat branches such as that offered by hawthorn, apple tree, hazel or spruce, or creepers like ivy and honeysuckle. The flimsy lattice of twigs is laid on a branch, not fitted – like a thrush's might be – into a secure crook. The male brings the material, but usually only the female builds. The twigs are interlaced so that the structure is quite rigid, but often the wickerwork is incompletely filled in so that even with a lining of rootlets or grass the two white eggs may be seen through the bottom of the nest. The nest is small for the size of the bird, and often forms a very insecure platform for the two squabs.

The wood pigeon's nest is, however, of average size among those of the *Columbidae* family. The ruddy quail dove *Geotrygon montana*, of Central and South America, builds a nest that is flimsy even for pigeons. Unusually, green leaves are worked into the structure, and both parents may continue to bring along fresh leaves during incubation (Gooders 1969–71). At the other extreme, the nests of the largest pigeons, the three crowned pigeons *Goura* spp, are large – about 18in (45cm) across – and substantially constructed from twigs lined with leaves. But the platform base is constant throughout the large family.

In the dense riverside vegetation of the Amazon basin the strange hoatzin *Opisthocomus hoatzin* builds a simple plat-

Right:
The European white stork ***Ciconia ciconia*** **builds a simple platform nest. A massive structure, to support these tall birds, it is often perched precariously on top of a chimney stack.** (*Brian Hawkes*)

Far right:
Long-tailed tit *Aegithalos caudatus.* (*Roger Hosking*)

form of woven twigs. Unlike the wood pigeon, whose chicks live for about four weeks on their nest, the downy young hoatzins very soon leave the nest and clamber around the nearby trees. They use their feet, bills and specially adapted wings which each have two 'claws' attached to the first and second digits. Very few other living birds have such claws, and only in the hoatzin chick are they used for climbing and gripping. The chicks' mobility, of course, means that they are less dependent on the simple nest platform, which must contribute to their chances of survival.

One of the best loved of all simple platform nests is that of the European white stork *Ciconia ciconia* which usually nests

Far left above:
Young barn owls ***Tyto alba*** **in their 'nest' – a haystack.** *(Roger Hosking)*

Far left below:
A swallow ***Hirundo rustica*** **feeding its young.** *(Roger Hosking)*

Left:
The purple heron ***Ardea purpurea*****, seen here with its young, prefers to nest on the ground among reeds and tall marsh grasses.** *(Dr K. J. Carlson ARPS)*

solitarily at traditional sites. Sadly this species has declined in recent times. The nest is a massive structure, as much as 6ft (2m) across. It is made of layer on layer of sticks and earth, with a shallow cup of finer materials such as grass and even rags and paper. Even the tall stork looks small on top of it. A nest is normally used year after year, with more material added each season. The storks feed in marshy grasslands but they have long nested close to man. In southern Europe favourite nest sites are old walls of towns, churches and ruins or farmhouse roofs; further north, chimneys, telegraph posts, trees and even specially erected stork poles are favoured. One nest on a tower in Thuringia (now part of East Germany) still occupied in 1930 was in use in 1549; a bill for its upkeep in 1593 still survives (Lack 1966).

The high regard felt for the bird today is nothing new. Near Celle in north Germany, I have seen a nest on the chimney of an old factory; it was considered more important to keep the storks than demolish the useless chimney. This love for storks is well recorded in Meindert de Jong's delightful story *The Wheel on the School*. It tells of a class of children who organise a project to bring back storks to their village because 'Storks on your roof bring all kinds of good luck . . . They build great big messy nests, sometimes right on your roof. But when they build a nest on the roof of a house, they bring good luck to that house and to the whole village that that house stands in.'

Herons and gulls build simple nests too, using materials which are typical of the site, which often appear to be thrown together. Eagles' nests are among the most studied of the simple platforms. All except the snake eagle build large nests, and most species build in trees. However, Verreaux's eagle *Aquila verreauxi*, in Africa, usually chooses a crag, as does the golden eagle *Aquila chrysaëtos* in Scotland. Nests on cliff ledges are invariably smaller than tree nests of the same species, probably because there is less space to build up the nest year after year. The largest nests are those built by the golden eagle. Its eyrie (as a raptor's nest is called) has often been found to

A golden eagle *Aquila chrysaëtos* brings a twig to its impressive nest-site. The eyries are usually built right at the top of Scots pine trees or on inaccessible cliff ledges. (*D. Platt/Aquila*)

A golden eagle adds a sprig of heather to the eyrie, either as decoration or reinforcement. The eaglet, having just been fed, is satiated. (*Niall Rankin/ Eric Hosking*)

measure 16ft (5m) from crown to base, and be well over a metre across at the top. One such nest was in a Scots pine tree and the top of the eyrie was actually the highest part of the tree (Seton Gordon 1955). A nest in British Columbia measured 21ft (nearly 7m) deep. A nest half as big as this has been estimated to contain more than a ton of sticks, accumulated over the years. In Africa, the crowned eagle *Stephanoaëtus coronatus* builds the largest nest; it may be 6ft (2m) across, and between 6 and 9 feet (2 and 3 metres) deep.

Both male and female eagles are builders. Dead sticks up to 1½in (4cm) thick are broken off, or picked up from the ground, large ones being carried in the feet, small ones in the bill. In mountain areas much heather may be collected, and, near the coast, seaweed. A new nest may take two or three months to

build (Brown and Amadon 1968). Within an eagle's territory there may be several eyries: a golden eagle usually has two or three, but from a single one to as many as ten have been recorded – the crowned eagle has only one (Brown 1970). These nests are used for many years, some individuals going to the same one every season, others using the nests in rotation. The reason for this, or why some individuals of some species are fonder of moving than others, is not known. Some authorities have suggested that the nest is so foul at the end of the season that it cannot be used next year. But many African eagles and golden eagles using only one nest in successive years breed successfully. The problem needs more critical study. The bateleur eagle *Terathopius ecaudatus* has been known to use the same nest for 18 years; crowned eagles for at least 20 years; and golden eagles for over 40 years. Indeed in Scotland, several eyries are believed to have been in use at the beginning of the century. Such favoured sites are used by successive generations for a pair of golden eagles need live only ten years to rear enough young to replace themselves.

The nests are renovated or rebuilt annually, according to their state of repair. Building activity reaches its peak just before laying but, in the golden eagle at any rate, it also goes on in summer and autumn; Seton Gordon recorded some activity in winter. Material is added even when the eaglets are in the nest, sometimes just a stick or two, at other times a substantial repair. A new eyrie may be established by a new pair, or by older birds to replace one blown down in a storm. Then the platform is made fresh with green leaves; the golden eagle often lines the shallow cup with the great woodrush *Luzula sylvatica*. Brown (1970) believes that all this 'compulsion to build may help to keep the female near the nest site, where she is needed, instead of going away to hunt, which could expose the eaglet to danger. This is, at any rate, a more likely reason for bringing green branches than aesthetic pleasure in adorning the nest . . . as sometimes suggested.' While Bannerman (1956) asks:

> Have birds an aesthetic sense? It would appear that the eagle has one. Why else should she carry to the nest, at intervals after the eaglets have hatched, rowan branches in full leaf, often with flowers on them? Why should she rarely, if ever, at this stage, bring the green-needled branches of fir which, earlier in the spring, she had intertwined to form a platform to hold the nesting cup? I have seen the mother eagle lay a rowan branch in delicate leaf gently across her eaglet's back, and then step back a pace to admire the effect.

The spoonbill's *Platalea leucorodia* nest is of sticks when built in a tree but of reeds if built on marsh ground. (*Roger Hosking*)

The turtle dove *Streptopelia turtur* builds a simple nest of twigs which make a fairly primitive platform but the song thrush (below) builds a more complicated, cupshaped nest with unusual hard lining. (*Dr C. J. F. Coombs*)

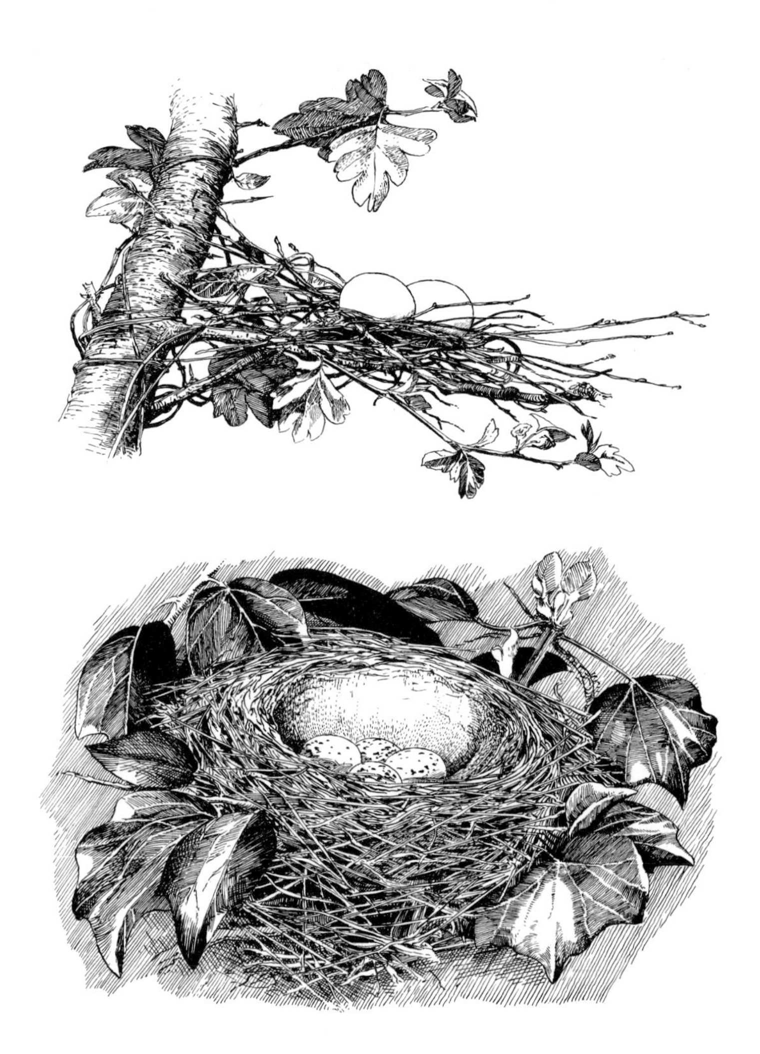

Who can fathom the mind of a bird? Certainly not those who give it credit for no intelligence.

Eagles are majestic birds. Having seen some one feels compelled to agree with Leslie Brown – the world's expert on eagles: 'To find the eagle's nest is to start to know eagles. It amazes me that so few seem to have made the attempt.'

4 Cupshaped nests

Cupshaped nests are to most people 'typical' nests. Ask a child to draw a nest and he will draw it cupshaped. Most are built by passerines; a huge variety of materials and sites are chosen, and size may be anything from a few centimetres in diameter to a sturdy great crow's nest.

A cupshaped nest is an optimum design. It can provide, for hundreds of species, such as warblers, the safest place for eggs

The paradise flycatcher *Terpsiphone viridis* often builds its cupshaped nest in a fork between a branch and twig. Here, in central Africa, this male keeps a watchful eye over the young. (*J. B. & S. Bottomley*)

and young. Many platform nests, even great eagles' eyries, have a hollow to hold the eggs – we have already noted that this is known as the 'cup'. One of the main developments in nest evolution has been the increasing importance of the cup, till for some birds the very nest itself is one complete cup with little or no sign of a simple platform. The nest's shape bears a direct relationship to the size of the builder for it is shaped by stereotyped building movements; as material is added to the base and the walls grow, the bird rotates, so pressing the inside wall into a smooth surface, often beautifully soft, especially after the lining has been added. The cup holds the eggs safely, its insulation helps to keep the temperature high for incubation, and it protects the helpless young.

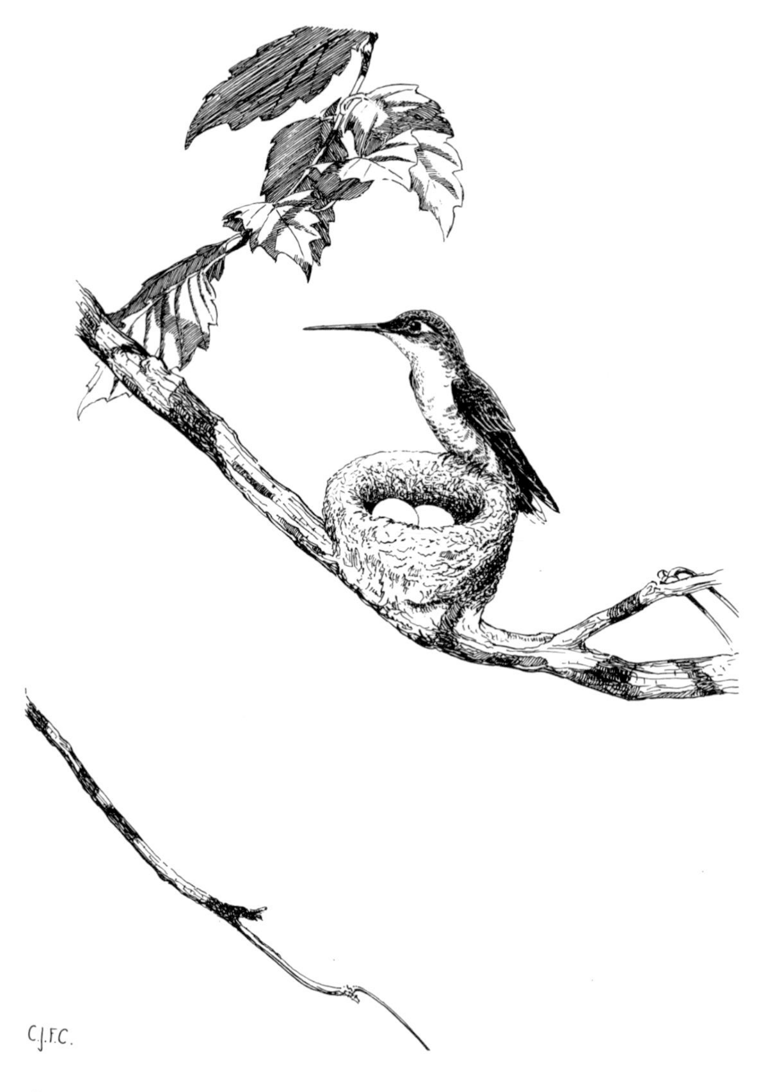

The ruby-throated hummingbird *Archilochus colubris* forms a perfect, tiny cupshape for its nest which it plasters on to a twig by using its adhesive saliva. (*Dr C. J. F. Coombs*)

Some of the smallest, daintiest and most beautiful of nests are made by hummingbirds. Equipped only with their slender bills, these tiny birds perform a most delicate and difficult task. They are very selective, going to great pains to get the 'right' materials, those habitually chosen by the species – the lightest and most filamentous – cottonwool, plant fibres, down, moss, lichen, hair, feathers, leaves, flower petals and seeds, and above all spiders' webs. Hummingbirds' nests are inconspicuous, being, not only so minute but also, wonderfully camouflaged and often cleverly hidden. The building of a nest, from the time the nest site is chosen, has been described in detail by Scheithauer (1967). The female brown Inca *Coeligena wilsoni*, from Colombia and Ecuador, is building with fine, long, clean threads of hemp, 4in (10cm) long.

> She placed these crosswise on the branch. She picked up the loose ends and while still flying, sometimes also perching, she coiled them in a clockwise direction around the branch. In this way a small firmly interlocking ball of material that looked like an encrusted cushion gradually came into being . . . The bird sometimes used its bill like a darning needle. Loose threads were stuck into the ball and drawn out at the side, as though properly stitched into place. It flew round and round the structure, removing loose threads and tucking them into other places. Thus it smoothed off and rounded the structure.
>
> When the cushion was large enough for the bird to lower herself onto it, she began working actively with her feet, twisting and turning round, and bending the edges of the cushion upwards with her bill and neck, thus making the structure into the shape of a flat saucer . . .
>
> The structure grew until, later, it was shaped like a hemisphere. Lying in the cup of the nest the bird now smoothed the edge carefully, using her neck, throat, bill and wings which were extended in a graceful movement that looked like an embrace. The outer wall of the nest received a final touch; she frequently used her tongue, sliding it quickly over protruding fibres until they lay flat.
>
> After five straight days of construction, the masterpiece was finished except for the internal fittings and the decoration and camouflage of the outside. These were completed during short flights, after the eggs had already been laid.

Many hummingbirds build a nest like the brown Inca's, tied to the upper surface of a branch or twig by threads of vegetable fibre or cobwebs – which are sticky and elastic and are indis-

A grey fantail *Rhipidura fulginosa* approaches its cupshaped nest of dried grasses with food for its brood. (*Gary Weber*)

pensable to many species in their construction work.

To many people 'hummingbird' means the widespread ruby-throated hummingbird *Archilochus colubris*, of the United States and Canada. This remarkable little long-distance migrant spreads an adhesive plaster of saliva on the twig chosen for its nest, and building upon the wafer thus formed it literally glues the nest to the support.

Variation in Ruby-throated Hummingbirds' Nests[1]

	NEST 1	NEST 2
Place	Cleveland, Ohio	Cleveland, Ohio
Time	—	15 June
Diameter of cup	0·94in (2·33cm)[2]	0·94in (2·33cm)
Depth of cup	0·69in (1·95cm)	0·81in (2·06cm)[3]
Outer diameter	1·43in (3·63cm)	1·69in (4·29cm)
Outer depth (avg)	1·20in (3·05cm)	1·31in (3·33cm)
Colour of nest	Dull brownish grey	Light grey
Materials	Fine plant down chiefly	Fine plant down, with pappus, seeds, bud scales, and two horse hairs smeared with saliva
Lining (inner wall)	The same	The same
Surface (outer wall)	Smooth, frescoed with bits of lichen, secured face up or down with spiders' silk	Smooth, with grey lichens secured with spiders' silk
Position	On twig	At branching of beech twig
Fixation	Nest material partly carried around stem. Saliva wafer[4]	By large wafer of saliva at base

Notes 1. Information from Herrick (1911). 2. Diameter at brim. 3. Probably too large; nest not quite perfect. 4. Nest detached when examined.

The ruby-throat's nest is often so perfectly modelled that it appears as if pressed in a mould, so true and even are its outer and inner walls and so perfect the rim of the cup. The only irregularity is seen at the base where the twig is enveloped. The curve of the inner wall of the cup is curiously bent inwards or overhung at the brim.

In the forests of the Mato Grosso in Brazil lives the fork-

A redwing *Turdus iliacus* brings mud to cement together its nest of leaves and grasses. (*J. B. & S. Bottomley*)

tailed wood-nymph *Thalurania furcata* which weighs about 4 grams (approximately one-seventh of an ounce) – less than Europe's smallest bird, the goldcrest *Regulus regulus*. Its finely woven open cup measures about 1½in (3¾cm) in internal diameter and the same in depth – narrower than a matchbox, but a little deeper. The smallest birds in the world are the bee hummingbird *Mellisuga helenae* and the vervain hummingbird *M. minima*. The former, confined to Cuba, weighs less than 2 grams and is about 2½in (6½cm) long. The vervain is 2¾in (7cm) long and is found on the islands of Jamaica and Hispaniola. Its nest, of silk cotton and lichen, is about the size of half a walnut shell. As Don Martino Springer, a Chilean Professor of Architecture, has said about a hummingbird's nest: 'Some of our present-day avant-garde architects could well profit by this example of art at its best, achieved by the greatest simplicity and not by orgies of form and colour!'

Hummingbirds are not alone in building well-camouflaged nests. In Africa the brubru *Nilaus afer*, a black and white shrike, builds a small, rather flat cup, usually quite high in a tree fork. It consists of small pieces of bark and bark fibres, smoothed down and made to blend with the branch. Even more wonderfully disguised is the nest of the Australian sitella *Neositta chrysoptera*. Fitting into a fork in a branch it is shaped like a deep cup and made of cobwebs, bark fibres and similarly workable vegetable matter. The whole of the outside of the cup is then decorated with flakes of bark from the same type of tree as the nest site so that from any distance the nest is almost invisible. Africa's *Terpsiphone* flycatchers build neat little cups which are graced by the spectacular presence of the beautiful, long-tailed bird.

In Europe many people agree that the most attractive cup-nest is that of the chaffinch *Fringilla coelebs*. 'Among all British birds none builds a more exquisite nest than a chaffinch – a small compact structure of felted moss, grasses, roots, wool or any soft substance, decorated externally with lichens or sometimes with birch-bark, fastened together by spiders' webs. A neat lining of hair and occasionally a feather or two complete this work of art which is built entirely by the female . . . in all manner of bushes or in hedges or the fork of a tree and is usually at no great height from the ground' (Bannerman, 1953). Curiously, lining is sometimes added after the hen has started to lay. Once I found a nest with an egg which had been covered by a feather and then pressed into the moss-felt of the nest wall, so being 'lost'.

The nests of the thrush family are substantial cups of leaves and grasses, often well cemented with mud. Those of the redwing *Turdus iliacus*, European blackbird *Turdus merula* and American robin *Turdus migratorius* are good examples. They build in bushes, in the fork of a tree, in hedge banks or even on the ground. The nests are four-part structures: a foundation of moss or twigs; a main structure of grasses, and sometimes more moss – I have often watched a female blackbird collect a beakful of dried grass and then carry it several metres to wipe it deliberately in soft mud, before adding it to the growing nest; a mud cup; and a lining of finer grasses.

In sheltered positions these strong nests last well into the following year. The moss often continues to grow and grass seeds germinate in the mud! Voles and mice use them as food stores, and sometimes other birds use them as a base for their nests. The song thrush *Turdus philomelos* is unique among British birds in leaving the cup with a hard lining. This is of mud, wet rotten wood, dung or peat cemented with saliva and moulded by the female's breast. The rim is often decorated with coloured seeds, especially those of the ivy which are stained pink from the juice of the berries. The female probably regurgitates these seeds while she incubates.

The nest of the carrion crow *Corvus corone* is sometimes used for several years in succession as in this third year cupshape which is built on two previous nests. (*Dr C. J. F. Coombs*)

Crows build the most substantial cupshaped nests, in a fork of a well-grown tree if available. The chough *Pyrrhocorax pyrrhocorax* of the Old World is exceptional: it chooses a ledge in a cave. Carrion crows *Corvus corone* in moorland districts have often to make do with stunted thorn, birch or alder, perhaps only 2 or 3m (6–10ft) above the ground. They usually build in a main fork or out on a limb, and rarely at the top of the tree as rooks *Corvus frugilegus* do. Twigs and small branches are first interwoven into a structure strong enough to withstand the gales of several winters; yet a nest is not used for two seasons in succession, although it may form the foundation of a new one. The twigs are strengthened and compacted still further with moss, earth and roots to form the basic cupshape. This is warmly lined with hair or wool, or both; sometimes the hair is plucked from the backs of grazing cattle or horses! The nest of a moorland crow may look coarse with jagged twigs, and as black as the bird from without; but within,

The chough's nest, although quite roughly made, is well-lined with the softest wool; it is usually built in a cave. (*E. V. Breeze Jones*)

This chough has built its nest in a derelict copper mine in Snowdonia, Wales. (*E. V. Breeze Jones*)

A Baltimore oriole *Icterus galbula* **feeds its young.** (*Donald Burgess/Ardea*)

the cup is thickly lined with white sheep's wool – it is hard to imagine anything cosier.

The great raven, *Corvus corax*, builds a similar structure but larger; a raven's nest is to a crow's as a basin is to a cup. Whereas a crow will usually build in a tree and occasionally on a cliff, the raven over most of its Holarctic range prefers a cliff ledge. Unusually among passerines, its nest often becomes an eyrie, used year after year. I know one site in a quarry which has been used regularly for over 20 years, another cliff site for 40 years, and D'Urban and Mathew (1895) knew a pair had built on cliffs in south-west England 'from time immemorial'.

5 Enclosed nests

People always think that a nest with a roof must be a safer 'home' for eggs or young birds than one without. Many birds throughout the world have evolved domed nests. As we shall see, they are not necessarily more secure than simple cup nests, but in tropical forests they do give safe refuge from the numerous marauders.

The leaf warblers *Phylloscopus* of the Old World provide examples of relatively simple domed constructions. The female

Far left above:
A pheasant-tailed jacana *Hydrophasianus chirurgus* **incubates its chocolate-brown eggs.** (*Peter Jackson/Bruce Coleman Ltd*)

Far left below:
Red-and-yellow barbet *Trachyphonus erythrocephalus* **at nest hole.** (*J. F. Reynolds*)

Left:
The willow warbler *Phylloscopus trochilus* **builds a fairly loosely-knit domed nest, usually on the ground.** (*E. V. Breeze Jones*)

usually builds the nest on or near the ground, of dead grass, leaves and moss – the various species have particular preferences. The chiffchaff *P. collybita*, for example, builds a rather loosely made domed nest 5–6in (12½–15cm) across, on the ground or a little above it, in woods or shrubby areas. It has a fairly wide side entrance – sometimes so wide that the nest appears only partly domed, so that all the brood can easily be seen. The nest is thickly lined with feathers – usually white, in my experience. The very closely related wood warbler *P. sibilatrix* lines her nest with fine bents and hair. This species' nest is almost always on the ground, often in a slight depression and built into the leaf litter on the floor of the wood so that only the entrance shows; though beautifully hidden from predators, it is susceptible to flooding in heavy rains. In a beech wood which has very little ground-cover other than last year's dead leaves the way the nest is hidden is a miracle of camouflage.

A large enclosed nest is that of the magpie *Pica pica*, found throughout much of North America, Europe and Asia. It builds its near-fortress of thorny or prickly sticks in the top of a tree or bush, lining it with mud and then with an inner softer lining of fine roots. Above the nest is an openwork, interlaced canopy of sticks, with a small inconspicuous entrance at the cup-edge. Many a schoolboy has climbed to rob a magpie's nest, only to wonder when he has finished his scramble where the entrance is. The nest is strong and lasts well, even in the exposed top of a gale-lashed tree. In later years the magpies may build anew on top of the remains, or a kestrel *Falco tinnunculus* (which like many raptors builds no nest of its own) may take it over.

The common wren *Troglodytes troglodytes* is one of the most familiar birds in the world, its tiny form and perky upright tail being affectionately regarded everywhere, but its nest is much less well known. An ideal wren nesting site would provide: suitable points to fix the nest material; a sheltered, snug position; concealment; and a clear flyaway from the entrance (Armstrong 1955). The nest is stoutly made of leaves, moss, grass and other plant material. The base is usually formed first, then the walls and the dome; lastly, working from the inside, the male (for he alone builds) fashions a slight overhang above the entrance which helps deflect raindrops. The threshold is specially strengthened with densely woven grasses. Sometimes the nest will be particularly well concealed, built of moss in a mossy bank, or of dead bracken amongst the same material. A wren has been reported to build a nest in a day, but three days would be more usual.

The magpie *Pica pica* constructs a strong and thorny base then encloses the nest under an openwork canopy of twigs. (*Dr C. J. F. Coombs*)

Sometimes the bird manages without all the ideal requirements, using a hole in a stone wall for example, or building among the roots in a hedge bank: here the natural crevice provides a roof, so it does not bother to build one!

A male wren is well known to build several nests, often 5 or 6, sometimes as many as 10 or 12. The Germans call these 'play nests', the British 'cock nests'. The male displays and sings before his chosen mate, and it is she who selects one of his nests and lines it with feathers, taking two to seven days to do so. The standard of this lining varies: a nest near a farmyard was crammed with chicken feathers, while a moorland nest I found had hardly any; up to 498 have been recorded. The cock wren will roost in one of his spare nests while the female is brooding in another.

But the common wren's well-ordered family system is surpassed by that of the cactus wren *Campylorhynchus brunneicapillus*, of Central America, and the south-western United States. Each young bird builds a nest of its own soon after it has become independent in the autumn, and this becomes a home throughout the coming year, giving protection from cold, rain and enemies. The flask-shaped nest is made of plant stems and grass, often in a cactus. The entrance is on one side, through a tunnel often 6in (15cm) long, although this tunnel may be omitted in a roosting nest. The nests may frequently be repaired until they become quite hard from dust and droppings (Gooders 1969–71). Several South American wrens build dormitory nests, which differ from the breeding nests. Cabanis's wren *Troglodytes modestus* constructs a breeding nest which is almost spherical with a noticeable cup

This globe-shaped nest, which is built by the male wren *Troglodytes troglodytes* may be one of up to a dozen of his efforts; the final selection is made by his mate. (*Eric Hosking FRPS*)

and protected entrance, whereas the dormitory nest is more flimsy, unlined and rather crudely made (Skutch 1940). The banded wren *Thryothorus pleurostictus* also builds a carefully protective breeding nest and many untidy roosting nests. The common wren *Troglodytes troglodytes* less frequently uses its nest as a roost through the winter and indeed the rain and winds soon destroy it – though one cock's nest I knew was used late in the summer, by a long-tailed fieldmouse *Apodemus sylvaticus*. This wren, though, like its neotropical counterparts, roosts communally in safe crannies and holes: as many as 46 of the little birds have been found in one nestbox, and 31 western winter wrens *T.t. pacificus* were once found in a nestbox 6in (15cm) square!

The dippers *Cinclidae* are fairly closely related to wrens and indeed their oval nests are rather like large wren's nests. Dippers *Cinclus cinclus* haunt fast-flowing streams and usually build beside the water. The large mossy globe, with a definite overhanging porch to shelter the entrance, is 'merely the shelter or envelope; below the median line, usually hidden beneath the lip, is the entrance to the real nest within, a cup of grass or sedge, nearly as large as the nest of a blackbird, lined with leaves' (Coward 1950). A traditional natural site is on a ledge in a bank, or stuffed in a crack by a waterfall – or even under it, the birds having to fly in and out through falling water! The nest is amazingly resilient; I have found one permanently soaking wet, being thoroughly splashed by a moorland torrent while the hen incubated four eggs. More amazing is the nest with young under an overhang on the bank of a stream, described by Alder (1963). After heavy overnight rain a flood had raised the level of the water by a metre and the torrent was washing the turf so that the nesting cavern was completely hidden. 'With remarkable "seamanship" the male was watched several times swimming and diving through the current with food for the nestlings. The rain stopped, the flood subsided, and a gap appeared below the turf, enough for the adult to scramble through.' Later the observer discovered that the nest had been lapped by water, but the young were alive, dry and warm.

The nest of Britain's long-tailed tit *Aegithalos caudatus* is one of the most beautifully designed and skilfully constructed in the world. Roger Hosking's photograph (opposite page 32) shows why it arouses wonder in anyone lucky enough to see it. Two sites may be distinguished: the favourite choice is in a bramble, gorse or honeysuckle bush, 3–16ft (1–5m) up; the second is high up, 22–65ft (7–20m) in the fork of a tree. It can be easy to see in the first case, if hard to reach through prickles and thorns, but very hard to spot in the latter, thanks to its camouflage. The nest is oval, even pear-shaped at times, with the round entrance well above the halfway line, usually near the top, occasionally right at the top. 'Felting is carried to its greatest perfection in this nest, in which shredded wool, green moss, spider-silk and lichens are artfully woven until a thick wall and dome surround the 5in [approx 13 cm] oval '(Coward 1950).

In March the two adult birds start to build carefully from the base up, the outer covering of grey lichen which is always such a distinctive feature of the finish being added as the nest grows. On average they take nine days to complete this stage, working mostly in the morning, and not much past noon; they may take as long as two weeks. Then after two or three days' rest, the pair take another 7 to 14 days to line the nest with feathers: only one feather is brought at a time, from as far as several hundred metres away. How so many feathers are

The female kestrel, bringing a vole to her young, has taken over a magpie's old nest built in a conifer. (*E. V. Breeze Jones*)

crammed in the nest is almost beyond belief. One nest I analysed was lined with 1,601, ranging from $\frac{1}{2}$–3$\frac{1}{2}$in (1–9cm) long (the bird is only 6in [14cm] long). Most were wood pigeons' feathers, but nine other species were involved too. A total of 1,000 feathers is not uncommon, the fewest I have heard of being 893, and the most 2,084 (Goodfellow 1973). The soft felted walls of the nest are elastic, changing shape to accommodate the growing chicks; usually a brood numbers 8 to 12, but there may be as many as 16!

Predation of nests is heavy. Owen (1945) found that 24 nests out of 70 examined were failures; and Gaston (1973) observed that only 9 pairs out of 36 managed to rear young successfully. Jays *Garrulus glandarius*, who were the commonest predators, ripped the nest to pieces, scattering the lining. In some the hole was enlarged, probably by a weasel or mice. In others there was no hole at all – mice had adapted the nest to make a home for themselves. A sad end to such hard work.

Many birds are named after their calls or colours. The long-tailed tit has more country or local names than almost any other bird; these are directly attributable to its nest: bottle jug, bottle tit, bottle tom, feather poke ('poke' meaning pocket).

The coconut-sized nests of the scrub-birds *Atrichornis* of Australia are noteworthy. The rufous scrub-bird *A. rufescens* is better known than its relative, the noisy scrub-bird *A. clamosus*, which was believed extinct until its rediscovery in 1961. Both build a loosely constructed, warbler-like nest of grass, leaves and small twigs, on the ground or a few centimetres above it in long grass or ferns. The lining is its special feature. It is made of a kind of wood pulp, chewed-up rotten wood which when dry has a surface like rough cardboard (Cayley 1950). As this lining takes some while to harden, nest-building may take about a month. The dry grass nest of the Australian double-barred finch *Steganopleura bichenovii* is more conventional, but the bird is more beautiful.

All the species so far mentioned in this chapter build some variations of the 'typical' domed nest. The three examples which follow are rather more eccentric. Firstly the Argentinian firewood gatherer *Anumbius annumbi*, a castle-builder of the spinetail group in the ovenbird family *Furnariidae*, is remarkable for its odd way of going about its building work as much as for the final result. For months after the young have fledged, the whole family of these amiable, sparrow-sized birds keeps together and uses the nest as a roosting place. The great nature-reporter W. H. Hudson described the nest-building thus:

> To build, the *Añumbi* makes choice of an isolated tree in an open situation, and prefers a dwarf tree with very scanty foliage; for small projecting twigs and leaves hinder the worker when carrying up sticks. This is a most laborious operation, as the sticks are large and the bird's flight is

The rufous-breasted castle builder *Synallaxis erythrothorax*, a close relative of the Argentinian firewood gatherer *Anumbius annumbi*, builds a complex and curious-looking nest, with two chambers and a connecting tunnel. (*Dr C. J. F. Coombs*)

feeble. If the tree is to its liking, it matters not how much exposed to the winds it may be, or how close to a human habitation, for the bird is utterly unconcerned at the presence of man. I have frequently seen a nest in a shrub or ornamental tree within ten yards of the main entrance to a house; and I have also seen several on the tall upright stakes of a horse-corral, and the birds working quietly, with a herd of half-wild horses rushing round the enclosure beneath them, pursued by the men with lassoes. The birds use large sticks for building and drop a great many; frequently as much fallen material as would fill a barrow lies under the tree. The fallen stick is not picked up again, as the bird could not rise vertically with its load, and is not intelligent enough, I suppose, to recover the fallen stick, and to carry it away 30 yards from the tree and then rise obliquely. It consequently goes far afield in quest of a fresh one, and having got one to its liking, carefully takes it up exactly by the middle, and carrying it like a balancing-pole returns to the nest, where if one end happens to hit a projecting twig, it drops like the first. The bird is not discouraged, but after a brief interview with its mate, flies cheerfully away to gather more wood.

The striking double-barred finch makes a domed construction of soft grasses and herbage. (*Gary Weber/Aquila*)

The hammerkop *Scopus umbretta* builds a vast domed structure of as many as 8,000 twigs and capable of supporting a man's weight. (*Dr C. J. F. Coombs*)

The nest of the *Añumbi* is about 2ft (60cm) in depth, and from 10–12in in diameter, and rests in an oblique position among the branches. The entrance is at the top, and a crooked or spiral passage-way leads down to the lower extremity where the breeding chamber is situated; this is lined with wool and soft grass, and five white eggs are laid (Recorded in Bowdler Sharpe 1898).

The ingenious simplicity of the crested swift's nest is noted in Chapter 6. The second eccentric nest-builder comes from within this family. The scissor-tailed swifts *Panyptila* of tropical America build an extremely elaborate nest. Hung from the underside of a branch or overhanging rock, it is a tube, from 7–24in ($17\frac{1}{2}$–60cm) long, with the entrance at the bottom. It widens at the top where a side pocket or shelf holds the eggs. Dry feathery tufts of plant seeds or birds' feathers are worked with saliva into a close felt which lasts well and may be used for several years, the birds merely adding a new shelf inside. In parts of central America these curious structures may be seen hanging under archways and from ceilings (Lack 1956).

Scientists still argue about whether the third example, the hammerkop *Scopus umbretta* of Africa, is more closely related to storks or herons. If its nest is anything to go by – neither. The nest is a massive, enclosed structure, quite unlike any other. It begins as a pile of sticks in a tree fork, or less commonly on the ground or on a cliff, but grows into a large inverted pyramid with a concave platform on top. The sides and back are built up, forming a deep cup with a V-shaped notch in front; this eventually becomes the entrance tunnel. The sides and back arch over, join and form a roof over a large cavity. More sticks, grass and mud are piled on the roof to make it waterproof. This roof may be a metre thick and has been known to support a man's weight without distortion. The nest chamber is 12–20in (30–50cm) in diameter and height, lined with sticks carefully woven so that no jagged ends project.

A Xhosa tribal legend has it that within the nest are three chambers – one for eating, one for hatching eggs, and a day-room. But although there are ledges and divisions inside, there is no systematic division into rooms. The entrance tunnel near the bottom on one side is 5–7in (13–18cm) in diameter, 15–24in (40–60cm) long, and is heavily plastered with mud to make it smooth. Some authors have remarked how materials are added after the nest is apparently complete, even after the eggs have hatched. The accumulation of old bones, rotting meat, hide, faeces, dead animals, paper and scraps has given rise to various native beliefs that to meddle with this nest brings disease and bad luck.

When the pair have finished, their handiwork measures as much as 6½ft (2m) in height, and 3–6ft (1–2m) in diameter; the largest nests may weigh several hundred kilos. An estimated 8,000 sticks and bunches of grass are used in the building. Such a construction takes a long time to complete. At one nest in Uganda, begun in early February, the roof was started on 22 February and completed by 27 February, the whole completed by mid-March, a total of six or seven weeks. A nest is sometimes used in subsequent years for breeding; one study-pair, on the other hand, roosted in an old nest during the construction of a new one (Kahl 1967). The nests are nearly impregnable, but in some areas barn owls *Tyto alba* have the temerity to take them over in the non-breeding season and prevent the rightful owners from using them.

6 Hanging and woven nests

Laymen and scientists alike have often expressed their admiration for the beautiful hanging nests built by many species. Whether simple cups suspended below a twig, or elaborate domed constructions like those of the weaver bird, they have an almost eerie mystique about them.

The goldcrest *Regulus regulus* of Eurasia, and the firecrest *Regulus ignicapillus* of Europe and North America, build tiny

The goldcrest *Regulus regulus* builds a small, penduline basket of soft moss, held by several 'handles' from a horizontal branch, usually of a conifer. (*Donald A. Smith/Aquila*)

baskets of moss suspended by several handles from a branch. They are sited typically under the tip of a horizontal conifer branch, the completed nest cup being only 1½in (4cm) in diameter. It is held together with spiders' webs and feather-lined.

Equally small and dainty are the nests of the white-eyes, tropical birds of the family *Zosteropidae*. Slung from two light twigs, up to 26ft (8m) high, they are made of lichen or moss, and sometimes tendrils and seedheads, the whole being held together by coarse cobweb and lined with plant down or hairs. Some have been found decorated externally with spiders' cocoons. Fragile to look at, they are strongly built although not easy to see, being so small and sometimes nearly transparent (Skead and Ranger 1958).

Several hummingbirds suspend cup-shaped nests from a twig or leaf tip. The many hermits of the genus *Phaethornis* are examples yet the tip of a palm leaf offers no support for perching. When a hermit starts to build she must work wholly on the wing, as Skutch (1973) describes.

> Bringing a weft of cobweb, she wraps it around the leaf tip by slowly floating around it on rapidly beating wings, one, two, or three times, always facing it. Then she fastens fragments of vegetation to the cobweb until she has a little shelf projecting from the inner side of the leaf tip, on which she rests while she shapes the mass into the form of a cup or concave bracket that will hold her eggs. The finished structure is often an inverted cone with a hollow at the top and a long 'tail' of fibres, shrivelled leaves, and other bits of vegetation dangling below the tip of the leaf.

Phaethornis preterei **a hermit suspends a minute cupshaped nest from a twig, adding a long streamer of vegetation as a counterweight.**
(*Dr C. J. F. Coombs*)

One of the most remarkable engineering feats by any bird is the hanging nest of the sooty-capped hermit hummingbird *Phaethornis augusti*, of Venezuela. This nest hangs by a stout cable of spiders' silk from an overhead support. But as the cable is attached to only one point on the rim of the open cup, the nest would tilt strongly if nothing else were done. Directly below the point of attachment the hermit fastens, with cobweb, little lumps of dry clay or pebbles which dangle below the nest and act as a counterweight to keep it level.

We have already seen that the Old World warblers build cup-shaped and domed nests; both these types have been adapted into hanging nests. The reed warbler *Acrocephalus scirpaceus*, for example, builds a deep, cylindrical-walled cup, the sides woven around two or more upright stems, as shown opposite page 84. It is firmly made of grass and flower heads, leaves and flowering stems of the *Phragmites* reeds in which the species so commonly breeds. A typical nest measures 2¾in (7cm) outside diameter, 3in (8cm) deep, and has a cup measuring 4cm across and deep.

The fan-tailed warbler *Cisticola juncidis*, of southern Europe, Africa and Asia, builds an unusual purse-shaped or elongated pear-shaped structure 4¾–6in (12–15cm) deep, with a small opening at the top; it is made by binding spiders' webs around grass or rush stems, the resulting cavity being lined with grass, flowers and plant down. The nest is well hidden, low down in the grass in which it hangs (Harrison 1975).

In Australia we may find 'a most wonderful example of a pensile nest near mountain courses'. The bird which makes it is the rock warbler *Origma rubricata*. Not a true warbler, it is also known as cataract bird, cave bird or rock robin, being never far from water in rocky ravines and gullies. In general shape its nest 'somewhat resembles a claret jug without a handle, having a long, slender neck and a globular and suddenly rounded bulb.' It has a side entrance, hooded over, and is composed of bark-fibre and grass, coated with fine green moss, cobwebs, even spiders' egg-bags, and lined with feathers. It is suspended by a cord of cobwebs from rock in sheltered places under overhangs, in caves, and even in culverts, drains and sheds. In a favoured site several may be built close together (Wood 1892 and Cayley 1950).

Members of the American family *Icteridae* and the Old World *Oriolidae* are famous for their hanging nests. The beautiful Baltimore oriole *Icterus galbula* builds a deep, purse-like structure, hanging from the tip of a branch, 8–10 metres above the ground. The female alone builds it, using strong, pliable plant fibres for the cup and fine grass and hair for

The fan-tailed warbler ***Cisticola juncidis*** **makes a pear-shaped nest by binding spiders' webs around plant stems then lining the inside with down.** (*Eric Hosking FRPS*)

lining. Herrick (1911) has described in great detail the building of one nest. It was first observed at 6.10pm on 14 May, on a pendent spray of elm. There were just a few 'grey bast fibres . . . wound to one twig only'. On 15 May work began at 7.05am. The nest grew as the bird, holding the fibres, worked with 'shuttle movements' of the bill, like a series of thrust-and-draw movements. With one thrust a fibre would be pushed through the mass; with a pull, that fibre or another would be drawn back and a stitch made. The motion was frequently too rapid to follow easily. In this work there was no deliberate tying of knots, but knots in plenty were tied all the time! There were many misses where threads remained loose and the weave was very irregular. By 1 o'clock on the 15th the nest was outlined in full. Work continued on the 16th, but was hampered by high

The white-throated warbler *Gerygone olivacea* builds a nest similar to that of a rock warbler, being shaped like 'a claret jug without a handle'. (*Gary Weber/Aquila*)

winds. On the 17th and 18th the bird was well hidden in the nest, and appeared only to be adding lining. The period of constructive work was thus about 4½ days. The nest was attached at six points by about 200 slender strands and was easily able to stand a strain of 8lb (about 3¾kg). Herrick believed that a typical oriole's nest might contain 10,000 'stitches', thousands of knots and loops; and have taken 40 working hours at five visits per hour, and 20,000 shuttle-movements to complete. No wonder that the species has been called America's greatest nest builder.

In Central and South America, two more groups of icterids, the oropendolas *Psarocolius* and the caciques *Cacicus*, also construct hanging nests, long pouches suspended from the tips of twigs. Some are as long as a metre with the entrance to the nest chamber at the very top. Strongly made of plant fibres, by the female alone, they hang quite freely, often at a great height, and their worst enemy is a high wind. Caciques and oropendolas are gregarious birds and up to 40 nests may be found in one tree.

The Baltimore oriole was described as 'stitching' its nest together; that was a figurative statement there, but stitches *are* really sewn by the next two species. Firstly the long-billed spider-hunter *Arachnothera robusta*, of south-east Asia. Madge (1970) has described a nest on the underside of a banana leaf, about 3 metres above a jungle track.

> It was built of coarse fibres which formed a tunnel with the opening towards the tip of the leaf and bulge towards the end where the nest cup of finer materials was situated. The fibres gave the appearance of being woven (indeed the nest looked something like half a Baya Weaver's nest, *Ploceus philippinus*, with a short, wide tube) but later examination showed that they were only pressed down. The nest was held in position with slings of spider's web, the ends of which were pushed into holes pricked on either side of the midrib. The little 'pimples' of web barely rose above the upper surface of the leaf and were only visible at close range. They were very securely fastened and I believe they were gummed into the holes, either by the natural stickiness of the web or by the bird's saliva. It was 17in [43cm] long and the entrance was $4\frac{1}{2} \times 4\frac{1}{2}$in [$11\frac{1}{2}$cm]. The nest cup was 4in long and $2\frac{1}{2}$in wide [$10 \times 6\frac{1}{4}$cm] and the part-circumference of the tunnel around it was $7\frac{1}{2}$in [19cm]. There were 81 points of attachment of slings on one side and 76 on the other. These were roughly in rows about 3in away from the midrib but getting closer towards the inside end.

Madge watched a second nest being built. When discovered, about a dozen spider-web slings were already in position, hanging down with a small quantity of fibres pushed loosely in among them. 'When the bird returned it hovered for several seconds below the leaf . . . then did a somersault, too quick for the eye to follow, and grasped the midrib of the leaf with its feet.' In this inverted position it made what appeared to be nest-building movements. At this time there were only 26 points of attachment, and the tunnel was but 2in (5cm) long, so the spider-web slings must be added a few at a time, and the tunnel filled in before a new length is started.

The second sewn nest is that of the well-known and much loved long-tailed tailor bird *Orthotomus sutorius*. This is

A golden oriole *Oriolus oriolus* suspends its nest from the cross forks of a branch. (*Dr K. J. Carlson ARPS*)

quite a common Indian bird, often found near human habitation, which explains its somewhat exaggerated reputation – the spider-hunter described above deserves to be just as well known. The hen forms a cradle by sewing leaves together with cottons, cobweb or silk from cocoons. The threads are pulled through small holes in the leaf which have been pierced by the bird's bill. The nest takes about four days to complete, with many false starts when threads break or the leaf tears. The nest cup itself fits snugly in the cradle and is made of cotton. The construction of the cradle naturally gives the nest more protection because it is disguised, and the leaf helps to repel tropical rain.

In south-east Europe and Asia, in bushy areas near water, lives the penduline tit *Remiz pendulinus*. It builds a remarkable nest, looking something like a long-tailed tit's (Chapter 5), but constructed and sited quite differently. Work is started by the male. He first twists plant down and grasses on to the tips of the twigs he has chosen as a nest-site, until a hoop is formed swinging 3–33ft (1–10m) above the ground. Then the base is thickened, and from there the bird works upwards from the inside; he is now joined by the female, and they toil away for about two weeks. The finished nest, whitish in colour due to the large amount of down from willow and poplar seed, is domed. But unlike the 'long-tail's', its entrance is a downward-slanting short tube, made last. The felted wall of the nest is very strong, but a gale may wrench the whole structure from its attachment to the tree.

The related kapok tit *Anthoscopus minutus* of south-west Africa suspends its domed nest from a small tree like an acacia. It has a tubed entrance 1–1½in (3–4cm) long just below the fixing point, but the tube is more horizontal than that in the penduline tit's nest; its opening is a mere slit, which may even be caused to close by a peck or two from the departing bird. Below the entrance is a noticeable larger opening, which in fact leads into a small dead-end. The purpose of this is disputed: some say it is to deceive predators such as snakes, others that it is used as a resting place for the non-incubating bird. The bird's name comes from the Afrikaans *kapokvoel*, because of the resemblance of the nest to the fruit and seed-hairs of the kapok tree (Gooders 1969–71).

The tiny verdin *Auriparus flaviceps* of the deserts of south-west America and Mexico builds a sphere 7¾in (20cm) high, which is large for a bird only 4in (10cm) long and weighing just over one-third of an ounce (less than 10g). It is 'so placed among the spiny branches of some bush that it is inaccessible without severe lacerations to any exploring hand reaching towards it' (Wetmore 1934), and what is more, it is cunningly made of twigs with their thorns projecting outwards! It is lined first with grass, then feathers. The same nest, or a new one serves as a winter roost.

Weaver birds' nests are renowned the world over. These are *the* builders in the bird world. 'Included in this family are some of the finest, most expert, and most famous of all avian architects. In no other single bird group of similar status has the habit of nest building been carried to greater heights, indeed their only rivals for excellence in this particular are some of the hang-nests or troupials (*Icteridae*) of the New World' (Friedmann 1949).

There are nearly 100 species of weavers. They fall readily into two groups: the genera *Ploceus*, *Malimbus*, *Nelicurvius*, *Notiospiza* and *Anaplectes* whose nests are unrivalled in the complication of their protective devices in construction and siting; and *Quelea*, *Foudia*, *Euplectes* and *Brachycope* which have simpler, comparatively vulnerable nests, typically built of and placed in herbage (Moreau 1960).

The shape of the nest depends somewhat on the nest site, on

whether it is slung between supports or suspended from the tip of a twig; and also on whether the weaver is a forest or grassland species. Yet within each species, the amount of individual variation in nest construction is incredibly small. Generally speaking weavers' nests are small globes with entrances at the side or below. The male does the building; the female may assist later with some lining to the egg chamber.

Nest building has been observed closely in a few species. In every case, bear in mind that the bird is doing nearly all the work with his bill, with a little help from his feet – craftsmanship of a high order. The method employed by the red-billed

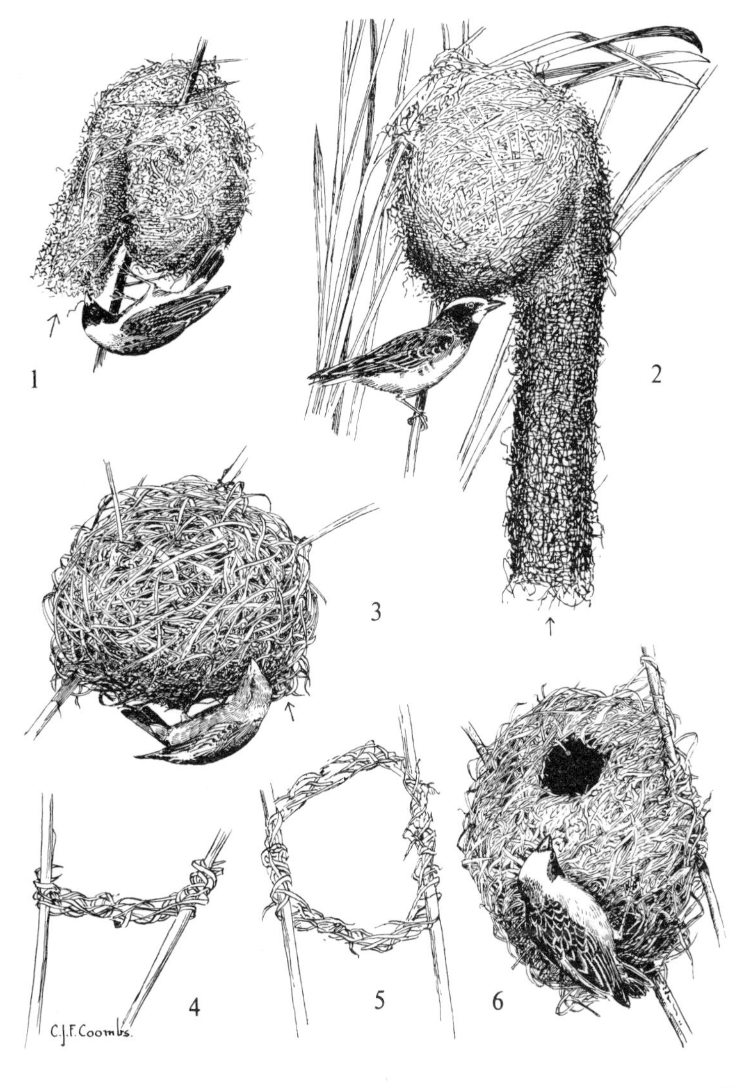

Different types of weavers' nests: (1) Retort as built by *Malimbus nitens*. (2) Retort with a long entrance tube as built by *Malimbus scutatus*. (3) Kidney-shaped. (4, 5 and 6) Stages in the construction of a globular nest built by a quelea. (*Dr C. J. F. Coombs*)

quelea or dioch *Quelea quelea* is typical. Its building technique is strictly ritualised in six stages:

1. Strips of grass are collected and knotted on to neighbouring vertical twigs to form two wads.
2. A firm, rounded initial ring is made for the two wads. This is the basic foundation of the nest and its equivalent can be found in all other weavers' nests. The builder perches on this ring to carry out most of the following building movements. The size of the nest depends on how far he can reach from his perch.
3. The walls and roof are built up.
4. The floor is established.
5. The wall and floor are thickened by additional weaving.
6. A porch is added, and the nest is complete.

Unlike the long-tailed tit which simply presses, and so felts, its materials together, the weavers do actually *weave*. Three principal stitches have been observed:

1. *Knotting* is employed in stage 1. A grass strand is held by the foot. The beak picks up a loose end, passes it round a twig, then releases it. The head is moved to the other side of the twig, the grass is pulled around the twig and inserted below the strand held by the foot. This is repeated and a genuine knot is formed.
2. The ring of stage 2 is built by *twining*. By this means material is added to existing fabric by stitching new strands right through the wad, and pulling the ends back through to near their starting points.
3. Stages 3–6 are constructed by *weaving*, by far the most complex stitch. A strand of grass is firmly fixed to one side of the initial ring by twining. The grass is then inserted through the edge of the 'roof-flange' until a loop appears on the far side. The loop is seized in the beak and the loose end pulled through taut. This action is repeated, again and again, until the strand is used up, the bird doing much of the construction from the same perching position at the bottom of the initial ring. As the back and walls grow it is necessary for the bird to hop incessantly back and forth stitching the strand.

The weaving is moulded into shape at all stages of the building by shaping movements, firstly with the head and beak which produce the curved form of the back and walls; and secondly, with the head and breast which mould the nest-cavity's final shape (Crook 1960).

Using these basic techniques, weaver birds construct four types of nests:

1. The *globular nest*, is a deep rounded basket with the entrance on one side at the top, so the bird enters horizontally

and descends to its eggs. Sometimes, like the dioch, it builds a small porch over the entrance. The walls may be thick and strong, or thin but nevertheless strongly woven and firmly attached to vegetation. Globular nests do not hang freely but are firmly bound to pairs of supporting stems, or many sticks which pass into the nest fabric.

2. In the *kidney-shaped nest* the entrance faces downwards so the birds fly up vertically into the antechamber and perch on the ledge before entering the egg-chamber. Nests of this type are normally 'slung' from a single twig orientated across the vertical axis of the nest, 'suspended' from two twigs or palm pinnae (one on each side of the nest), or occasionally 'pendent', the nest being attached at the top to a single downhanging twig or palm leaf. These nests frequently have an inner lining to the roof and egg-chamber, added by the female.

3. *Retort-shaped nests* with a funnel entrance are made only by the blue-billed weaver *Malimbus nitens*. The nest has a widely spreading entrance leading diagonally downwards from the egg-chamber. It is loosely constructed and is usually slung or suspended over water.

4. There are many *retort-shaped nests with tunnel entrances*. These are characterised by elongated entrances projecting diagonally or vertically down from the initial ring. The longest tunnels always hang vertically. Nests like these are commonest in arboreal forest species and may be slung or suspended. The entrance tube of the village weaver *Malimbus scutatus* may be as long as 2ft (60cm). Birds entering these nests 'do a remarkable, aerobatic sweeping-down to the entrance tubes and then "diving" upwards into them with closed wings, the momentum carrying the birds to the top of the tube, and the sill of the nest without the fabric being touched' (Crook 1960). The texture of the fabric of this particular nest is especially neat. Although the walls are thick and firm, the tube of criss-cross fibres is like a transparent net – a bird can clearly be seen inside (Crook 1963).

Weavers' nests are intricately woven, but can be made quickly. One kidney-shaped nest was observed being started at 8.30am; by 9.45am the initial ring was complete; by 1.30pm the basket was present but thin and incomplete; at 3.15pm courtship of a female began; by 5.30pm the building was nearly done; the following morning the rest of the lining was added and the nest was finished! Retort-shaped nests may take up to a week to build, especially those which are finely woven and where robbery of nest-materials by neighbours is a continual hazard.

The red-headed weaver *Malimbus rubriceps* is unique in building its nest of twigs. These cannot be woven as described

The yellow-faced honeyeater *Meliphaga chrysops* of Australia, seen here with expectant brood, builds a nest similar to that of an oriole. (*Gary Weber/Aquila*)

above. The male chooses a twig, strips the leaves off it, and breaks it off the tree. Then, quite remarkably, he nicks the bark with his bill so that at least one tag of bark hangs free. *All* twigs are built into the nest by knotting these tags of bark into other twigs. A ring is formed first, then the pendant retort-shape is built up by knotting twigs together vertically. As the nest nears completion twigs are tied on horizontally too. The nest is lined with leaves and the whole fabric is remarkably strong (Crook 1963).

All these wonderful nests have evolved, by natural selection, into shapes which have considerable survival value. Some are made of specially strong fabric to withstand the wind in exposed sites. They may be streamlined too. Kidney-shaped nests in areas of high rainfall have a roofing lining to prevent rain penetrating the egg-chamber. The threshold of the egg-chamber is so high that it is almost impossible for eggs to roll out when the nest sways. The funnels and tubes of retort-shaped nesst are helpful in protecting eggs and young from predators such as tree snakes and birds, although there are appreciable losses to the African harrier hawk *Gymnogenys typicus* and the didric cuckoo *Chrysococcyx caprius*. We must agree with the 1831 comment of Professor J. Rennie: 'The whole fabric of the nest is most ingeniously and elegantly woven; and the wonderful instinct of foresight (or whatever else we may choose to call it) displayed by the little architect in its construction, is calculated to excite the highest admiration'.

However, some tropical swifts may be said to win the prize for the most extraordinary hanging nests. The African palm swift *Cypsiurus parvus* builds a small pad of feathers which it glues with its saliva to the underside of a hanging palm leaf. The lower edge has a small rim to support the one or two eggs, but as this is insufficient to hold them properly they are glued with saliva into place. They cannot be turned over during incubation, as most eggs are, but the species nevertheless breeds successfully. Because the nest is hanging vertically the adults incubate by clinging to the nest with their toes, squatting bolt upright. In high winds nest, eggs and bird may be blown upside-down (Moreau 1941a).

The crested swifts *Hemiprocne* of south and south-east Asia have in relation to their size what are perhaps the smallest nests of any bird. Living on the forest edge and in clearings, they build high in large trees. All the species have a similar nest of a few small feathers and thin flakes of bark, probably collected in flight, glued together with saliva to form a tiny shallow cup, just big enough to hold one egg. That of the Indian tree swift *H. coronatus* is typical of the group: it is hemispherical, flimsy, about 1½in (4cm) long, 1in (2½cm) wide, and about ⅓in (1cm) deep. The egg measures 23½ × 17mm. The whiskered tree swift's *H. comata* has been reported as even smaller, just over 1in (3cm) long (Bowdler Sharpe 1898). These pigmy nests are glued against the *side* of a bare lofty twig, ½–1in thick so that the top of the nest is level with the top of the twig. It is impossible for the incubating bird to sit on the tiny structure, so it perches on the twig, facing away from the nest, just covering the egg with its brood patch. Both sexes incubate and the relieving bird shuffles sideways on to the egg as its mate sidesteps off (Phillips 1961, Tweedie 1960, Austin 1961).

The young palm swifts and crested swifts are in the nest at least ten days less than common swifts *Apus apus*, which is no doubt an evolutionary advantage to compensate for the exposed nesting site; other swifts nest in sheltered places, such as caves or under the eaves of houses.

7 Mud nests

So far we have traced the development of the nest from the humble platform of the dove to the elaborate architecture of the weaver. All these birds use mainly 'traditional' materials – sticks, grass and moss with softer materials for lining. They are builders, weavers and tailors: now we come to the masons, the birds that build almost exclusively with mud.

No family shows how to use this material better than the swallows and martins *Hirundinidae*. Through the centuries many species of these delightful birds have nested in close

The crag martin *Hirundo rupestris* usually makes its nest in rocky terrain on the arid slopes of low southern Palearctic mountains. (*Dr K. J. Carlson ARPS*)

association with man, happily using his buildings instead of the crags, cliffs and caves of natural nest-sites.

Hirundine nests may be cup-shaped or retort-shaped. The swallow *Hirundo rustica* of the Holarctic builds inside a darkened building, such as a barn, garage or shed. The adults are adept at diving at high speed through open doorways, broken windows, or small gaps in weather-boarding to gain access to a suitable 'cave'. The nest built by both adults is of pellets of mud, reinforced with bits of herbage and lined with grasses and a few feathers. Usually it is attached to a beam in the roof, sometimes close against a wall when it may be more like a half-cup. The crag martin *Hirundo rupestris* of the arid slopes of the low mountains in the southern Palaearctic builds a similar nest, but without the grass binding material which is such a feature of the swallow's, and almost always in natural rocky sites.

The house martin *Delichon urbica* is most well known where it chooses to fix its nest on the walls of houses, close up beneath the eaves so that the chamber is quite enclosed save for a small entrance hole in the rim. In many parts of Europe and northern Asia it remains a bird of open country, the nest being stuck to a rock or cliff under an overhang, but in Britain this natural site is now rarely used; I have seen cliff nests there only once, on the Great Orme in north Wales, a site which has been recorded since 1901 (Jones and Dare 1976). In sheltered situations the nests last well and need only be patched up next year, appearing multicoloured when the adults use a different mud from that used in an earlier building session. The account of martins' nest-building written in 1773 by the famous naturalist Gilbert White, to his friend Thomas Pennant, can hardly be bettered:

> About the middle of May, if the weather be fine, the martin begins to think in earnest of providing a mansion for its family. The crust or shell of this nest seems to be formed of such dirt or loam as comes most readily to hand, and is tempered and wrought together with little bits of broken straws to render it tough and tenacious. As this bird often builds against a perpendicular wall without any projecting ledge under, it requires its utmost efforts to get the first foundation firmly fixed, so that it may safely carry the superstructure. On this occasion the bird not only clings with its claws, but partly supports itself by strongly inclining its tail against the wall, making that a fulcrum; and thus steadied, it works and plasters the materials into the face of the brick or stone. But then, that this work may not, while it is soft and green, pull itself down by its own weight, the provi-

The house martin begins to plaster the mud to a surface, usually the wall of a house or barn, close up under the eaves. (*Frank V. Blackburn*)

The final touch to the martins' nest is the lining, either of feathers, hair and wool, or grasses, straw and moss. (*W. S. Paton/Aquila*)

dent architect has prudence and forbearance enough not to advance her work too fast; but by building only in the morning, and by dedicating the rest of the day to food and amusement, gives it sufficient time to dry and harden. About half-an-inch seems to be a sufficient layer for a day. Thus careful workmen, when they build mud-walls (informed at first perhaps by this little bird), raise but a moderate layer at a time, and then desist, lest the work should become top-heavy, and so be ruined by its own weight. By this method in about ten or twelve days is formed a hemispheric nest with a small aperture towards the top, strong, compact, and warm; and perfectly fitted for all the purposes for which it was intended. But then nothing is more common than for the house sparrow, as soon as the shell is finished, to seize on it as its own, to eject the owner, and to line it after its own manner.

After so much labour is bestowed in erecting a mansion, as Nature seldom works in vain, martins will breed on for several years together in the same nest, where it happens to be well sheltered and secure from the injuries of weather. The shell or crust of the nest is a sort of rustic work, full of knobs and protuberances on the outside; nor is the inside of those that I have examined smoothed with any exactness at all; but is rendered soft and warm, and fit for incubation, by a lining of small straws, grasses and feathers, and sometimes by a bed of moss interwoven with wool. In this nest they tread or engender, frequently during the time of building; and the hen lays from three to five white eggs . . .

They are often capricious in fixing on a nesting place, beginning many edifices, and leaving them unfinished; but when once a nest is completed in a sheltered place, it serves for several seasons. Those which breed in a ready finished house get the start in hatching of those that build new by ten days or a fortnight. These industrious artificers are at their labours in the long days before four in the morning. When they fix their materials they plaster them on with their chins, moving their heads with a quick vibratory motion. They dip and wash as they fly sometimes in very hot weather; but not so frequently as swallows. It has been observed that martins usually build to a north-east or north-west aspect, that the heat of the sun may not crack and destroy their nests; but instances are also remembered where they bred for many years in vast abundance in a hot stifled inn-yard against a wall facing to the south.

Many modern house-holders are less kindly disposed to

house martins because the droppings of the young soon foul the surrounding area. Some even poke down nests, others hang 'scarecrows' on the wall to deter the birds, but despite this the martins will often persist in attempting to nest there.

Some house martins have been seen building a short spout to protect the entrance of the nest. This evolutionary development has been taken to extremes by the red-rumped swallow *Hirundo daurica* of Africa and Asia, and the cliff swallows *Petrochelidon*. The red-rumped swallow builds a half-cup like a house martin, or a cup of mud pellets stuck on a ceiling of rock scantily lined with feathers, with the entrance stretching like a spout along the rock surface. This spout is 2–3in (5–8cm) long, and may be extended even further in a drooping tube up to 6in (15cm) long. Nests in the cold Tibetan highlands have markedly thicker mud walls and thicker inner linings of sheep's wool, yak hair and feathers (Voous 1960).

The cliff swallow *P. pyrrhonota* of north America builds a fascinating mud-pellet cup with a tubular entrance. The fairy martin *P. ariel* of Australia builds an equally well constructed

A blue-faced honeyeater *Entomyzon cyanotis*, which has taken over a magpie lark's nest, feeds it young. (*Gary Weber*)

nest which gives rise to the bird's country name of bottle swallow (Cayley 1950). It has been calculated that its nest is made of about 1,300 pellets of mud, measures 6×4in (15×10cm), with the downward and outward projecting tube-entrance being up to 1½in (4cm) long, and weighs about 1½lb or nearly ¾kg.

Another bird which has a nickname because of its nest is the magpie-lark *Grallina cyanoleuca* which to many Australians is the 'mudlark'. Its beautifully proportioned bowl-shaped nest is built on a bare horizontal limb of a tree, usually one which is near or overhanging water. It is composed of mud, strengthened like ancient Egyptian bricks with small sticks, grasses, feathers, horse-hair and even fur, and lined with fine grasses. Though weighty, it is very firmly fixed and rarely comes to grief. It has been described as looking 'very like an exceedingly rude and ill baked earthenware vessel, just such an one indeed, as Robinson Crusoe manufactured on his island' (Wood 1892).

The nest of another black and white bird, willie wagtail *Rhipidura leucophrys*, is often built in the same tree on another horizontal branch. The two species quarrel but little, and probably assist each other in spotting predators. Both are conspicuous birds, neither hides nor disguises its nest, and so four pairs of eyes instead of two are an advantage in the defence of their broods.

The apostle bird *Struthidea cinerea* and the white-winged chough *Corcorax melanorhamphus* are thought to be related to the magpie-larks. Both are found in Australian open forest and build reinforced mud nests, weighing up to 5lb (2¼kg), on a horizontal limb up to 50ft or so (15m) above the ground. The species are communal, living usually in small flocks, and remarkably all members of the flock – which may be a dozen or more – help to build the nest and rear the young. Sometimes more than one female may lay in the one nest, but this has no real advantage because the mud is quite unable to stretch to hold an abnormally large family.

In the great forests of west Africa live the bald crows *Picathartes*. They are strange birds, probably related to the babblers *Timaliinae* of Asia and Australia. Although apparently dwelling deep in forests, they actually spend most of their lives in or near caves, where they build a cup of mud like a large swallow's nest. It is strengthened with grass, lined with vegetable matter, and may last for years, with repairing and relining. Several pairs usually nest in the same cave, tolerating each other quite contentedly, even perching unconcernedly on each other's nests. Some nests may be near the cave entrance, but others are deep inside in the dark: that, coupled with the

The greater flamingo *Phoenicopterus ruber* builds a circular mound of mud then lays its egg in the depression on the top. (*G. J. Broekhuysen/Ardea*)

fact that they are usually made of mud from within the cave, means that they are very well camouflaged on the rocky wall (Brosset 1969).

Where conditions are suitable, both the greater flamingo *Phoenicopterus ruber* and the lesser flamingo *P. minor* build mud nests. Both are roughly circular mounds, about 13¾in (35cm) across, broadest at the base, with a shallow depression at the top. Sometimes the nest has little or no height, being merely a circular bulwark of mud with the impression of the bird's legs distinctly marked on it. Others are 6–7¾in (15–20cm) tall. Men used to be puzzled as to how this tall bird could sit on such a small nest: it was seriously thought that the hen straddled the nest! But in fact she sits with her legs tucked beneath her like most other birds.

At the other end of the size scale is the spotted morning warbler *Cichladusa guttata* of east Africa. It builds a nest almost entirely of mud with a few fibres as lining, the deep cup always being plastered onto a bough or other support. Compared with other members of its family, this bird uses a most curious nest-material (Mackworth-Praed et al. 1973).

In south-east Europe and Asia Minor lives the rock nuthatch *Sitta neumayer*. Further east is the closely related *S. tephronata*. Both are small birds of rocky habitats such as limestone karst country. Each is only $5\frac{1}{2}$in (14cm) long but they build massive nests. Mud is plastered by both sexes into a crevice in the cliff, or attached to a steep face of rock, or wedged under a rocky projection, and a bottle-shape is built on to this base. A funnel-shaped entrance is added, the hole just big enough to admit the adult birds. The mud walls are about $\frac{3}{4}$in (2cm) thick and

The spotted morning warbler ***Cichladusa guttata*** **of East Africa plasters its mud cup onto a tree; occasionally the nest will be lined with a few fibres.** (*J. F. Reynolds*)

Far right:
A reed warbler ***Acrocephalus scirpaceus*** **feeds its young.** (*Roger Hosking*)

inside is a substantial felted cup of moss, hair and feathers – especially those of the rock dove *Columba livia* (Voous 1960). The whole nest may be 7¾–9¾in (20–25cm) across and weigh over 77lb (35kg), about 1,000 times the weight of the bird! The eastern rock nuthatches in particular decorate the outside of the mud, while it is wet, with indentations of their bills, with feathers and wingcases from beetles. The whole nest may take ten days to build (Harrison 1975). The common nuthatch *Sitta europaea* on the other hand, simply plasters up a tree hole until it is the right size.

Writers a century ago believed that among 'mason' birds one was pre-eminent: it had no equal, and many said no second. This was the ovenbird. There are in fact two closely

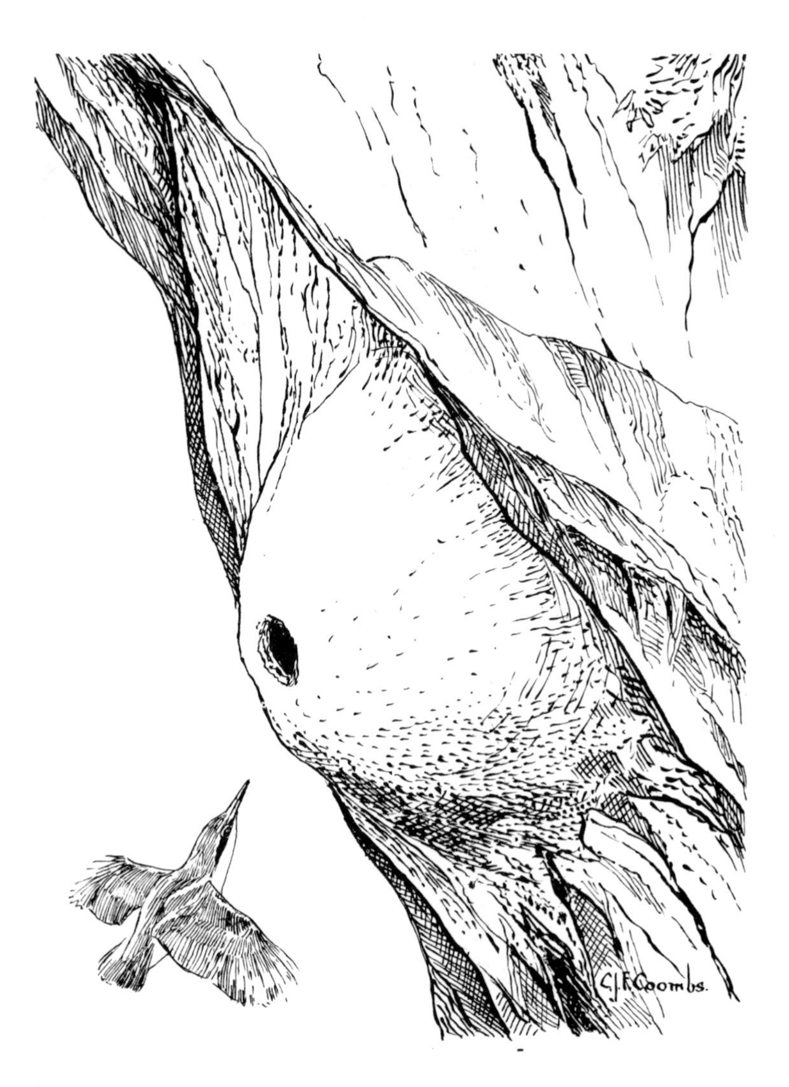

The rock nuthatch ***Sitta neumayer*****, although a small bird, builds a huge mud nest of up to 1,000 times its own weight, usually in rocky terrain.** (*Dr C. J. F. Coombs*)

related species, the rufous ovenbird *Furnarius rufus* and the crested ovenbird *F. cristatus*, whose breeding habits are almost identical. Both are found on the grassy plains of southern South America and are lively birds about the size of a thrush. The nests are made of mud mixed with grass and hair, kneaded with the beak to a kind of cement which, when dry, is extraordinarily hard. Durrell (1956) in Argentina described one thus:

> We came upon a tree-stump in the long grass, and perched on top of it was an ovenbird's nest. I was amazed, when I examined it closely, that a bird of this size could produce such a large and complicated structure. The nest was globe-shaped, roughly twice the size of a football, strongly made of mud combined with roots and fibres, so that it formed a sort of avian reinforced concrete. Looked at from the front, where there was an arched entrance, the whole thing resembled a miniature version of an old-fashioned bread-oven. I was interested to see what the inside of the nest was like, so,

Far left above:
A Bradfield's hornbill *Tockus bradfieldi* perches above its nest hole. (*Peter Steyn/Ardea*)

Far left below:
A crowned plover *Vanellus coronatus* lays its egg in a scrape in the ground. (*Norman Myers/Bruce Coleman Ltd*)

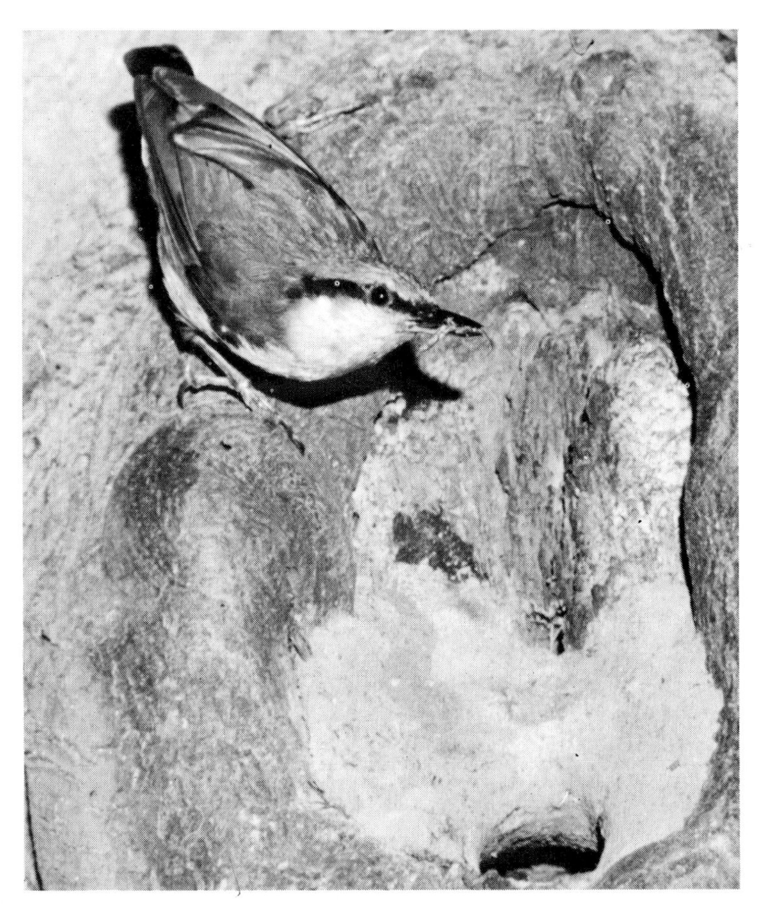

Left:
A nuthatch *Sitta europaea* plasters a natural hole until it is just big enough to admit the adult bird at a squeeze. (*E. A. Janes/NHPA*)

being assured that it was an old one, I prised it off the tree-stump and cut carefully through the brick-like top of the dome with a sharp knife.

When the top was removed, the whole thing looked like the inside of a snail shell: a passage-way ran into the left for some six inches from the arched door, following the curve of the outside wall, but bent in at the right of the door so as to form the passage-way. Where the passage ended, the natural shape of the nest formed a circular room, which was neatly lined with grass and a few feathers. While the outside of the whole structure was rough and uneven, the inside of the little room and the passage-way was smooth and almost polished. The more I examined the nest, the more astonished I became that a bird, using only its beak as a tool, could have achieved such a building triumph. No wonder the people of Argentina look with affection upon this sprightly bird that paces so pompously about their gardens and makes the air shiver with its cheerful, ringing cries.

Both birds of the pair set to work on the construction; one may collect mud and the other knead it into shape on the chosen nest-site. A wide variety of supports has been recorded – tree branches, telegraph poles, fence posts and buildings. The nests vary somewhat in size, consisting of 7¾–11lb (3½–5kg) of mud. They may be finished in four or five days if mud is readily available, but dry weather may delay the builders or even cause them to give up and start a new nest later. Two broods are reared in the same nest each year, and the whole family will roost in the 'oven' in bad weather.

As with many hole-dwelling species, when the nest is finally deserted swallows and sparrows take it over (Andrade 1969).

It is not surprising that such a spectacular, conspicuous nest has given rise to much speculation and superstition. Perhaps the most interesting belief is that of some natives who say that God sent the ovenbird to men to show them how to build solid homes.

8 Ground nesting

It might be argued that nest construction reaches the heights of ingenuity and wonder in the weaver birds' woven fabric and the ovenbird's mud ball, and that all others are examples of simple constructions, cup-shaped or enclosed nests. Many species, such as gulls, waders, warblers and grouse, breed on the ground, and even ospreys *Pandion haliaëtus* will build an eyrie at ground level when their territory is a treeless, low-lying shore. Some of these nests have been described already but it needs to be emphasised that many species, both passerines and non-passerines, do not nest in bushes and trees. The next three chapters, therefore, emphasise where birds nest.

Black-headed gulls *Larus ridibundus*, like others in their family, are colonial and nest on the ground. (*William S. Paton*)

Birds which walk awkwardly and have long wings, like the wandering albatross *Diomedea exulans*, one would assume must nest on the ground. But the equally aerial, magnificent frigate-bird *Fregata magnificens* builds a simple platform in tropical trees. It will occasionally crash-land beneath the trees when there is not enough wind for it to fly properly; and there it dies, so ill-equipped is it for life on land. More practically, the wandering albatross, in treeless South Georgia, builds a considerable structure on the ground. Both male and female first scrape a circular trench with their beaks, piling the soil, moss and tussock grass in the centre. They add more to the heap, and trample it down into a solid, peat-like mass. This circular nest is about 3ft (1m) across, 1–3ft high, and has a shallow bowl at the top for the single egg. It is 'shaped like a miniature volcano' (Jameson 1958) and a deserted nest is reported to make a remarkably convenient seat! Old nests are often repaired and used again. They are built high to keep the top of the nest clear of the winter snow: the birds' breeding cycle is so long that young are in the nest throughout the winter, protected from the elements by thick fat and down.

Swans are even larger birds with proportionately larger nests, made of heaps of plant material with a hollow at the top to hold the eggs. The mute swan *Cygnus olor* is well-known wild or as an introduced species in many parts of the world. It nests in remote swamps and river systems, and also by ponds in city parks. Its nest is 3–4ft (1–1¼m) across and usually tall enough to be above any normal rise in water level. It is reported that both male and female can anticipate flooding and raise the level of their nest accordingly. The cob collects new material and piles it on the side of the nest, and it is roughly laid out by the pen and flattened by the crown of her head. The eggs are then rolled on to the higher surface (Bannerman 1957). The biggest swan in the world, the trumpeter *Olor buccinator*, is also the rarest. During the nineteenth century thousands were slaughtered for their beautiful plumage to make down coverings and powder-puffs. Thanks to the Migratory Bird Treaty Act of 1918 which protected the birds throughout their North American flyway, a bird which was nearly extinct now numbers about 1,500, about half that number being in the United States, and half in Canada. Truslow (1960) discovered one nest in Montana, in a little bulrush marsh. It was built on a muskrat lodge 2ft above water, and measured about 5ft in diameter. The method the swans used to build the nest was interesting. Grabbing a big billful of vegetable material, the trumpeter pivoted 180° and threw the load towards the nest site. Circling the nest and continuing the

The eider duck *Somateria mollissima* lines her nest thickly with her soft breast down in order to help insulate and protect her eggs. (*Dr K. J. Carlson ARPS*)

operation the swans piled up the material yet kept movement to a minimum. This was a perfect example of 'backward-throwing', or 'the steam-shovel method' as Truslow called it.

Most, but not all, wildfowl nest on the ground, the duck building a simple construction of grass in a scrape, lined with down and feathers from her own plumage. This is usually in thick herbage near water, often in heather or long grass, under a bush or in willow scrub. Sometimes in tussocks of grass the nest appears to be in a hollow with grass stems arching over to form a natural roof. The eider *Somateria mollissima* which breeds along the shores of Arctic Europe, Asia and America, south to about 55°N on European coasts and 40°N on the western shores of the Atlantic, is one of the most numerous ducks in the world; approximately 2,000,000 winter in Europe alone (Sharrock 1976). The sitting duck's cryptic plumage helps protect the nest, but in open tundra and coastal marram grass even that is not enough to prevent considerable predation by crows, skuas, gulls and foxes. Of all the ducks, the eider lines

The sociable Canada goose *Branta canadensis* also lines its nest thickly with down and grass. (*William S. Paton*)

her nest the most profusely with her own down. Firstly, she tramples a nest site in the grass; grass, heather or seaweed are picked to line the nest and cover the first eggs. Down is plucked from the breast and flanks, from the laying of the third egg onwards, and added till the whole clutch (usually four to six eggs) is thickly and softly embedded in it. The duck then broods alone for 26–28 days, rarely leaving the nest and then only for a hasty drink. The down not only conceals the eggs when the duck leaves the nest but also helps to insulate them from the cold damp ground and sharp northern air.

Man has long appreciated the softness of eider-down (its Latin name is *mollissima*, 'softest'). Although sometimes nesting solitarily, eiders are social, even colonial in places. This has been exploited in Scandinavia, Iceland and parts of Greenland. The polar eskimos of the Thule district have been very careful with the conservation of their duck colonies, unlike many others who just hunt the birds for eggs, meat, skin and feathers (Freuchen and Salomonsen 1959). The most organised 'farming' of the down has for centuries been found in Iceland, and in times past was practised by monks in Scotland, and on islands off the north coast of England and in the White Sea. The down is gathered once or twice from each nest – immediately after the laying of the eggs and sometimes halfway through the incubation. Up to ½lb (225g) of down has been collected from one nest, but less than 1oz (28g) from nests harvested once already.

Some Icelandic colonies may have 6,000 nests (Ogilvie 1975), and although the birds are wild and free-flying they show many signs of domestication, even allowing visitors to stroke them. The nesting area is decorated with flags, bunting, little windmills, even musical instruments, and artificial, semi-detached nest sites. All this, it is claimed, encourages the birds to breed and prevents predation! The value of the down from these colonies is of considerable importance to the Icelandic economy, only two other nests being of similar importance – the guano cormorants' and the edible nests of Asian swiftlets. There are many artificial substitutes today, but eider-down is still considered by many to be the best filler for sleeping bags, pillows and bed-covers, although the famous French naturalist, Buffon, in the eighteenth century remarked tartly that the hardy hunter clothed in bearskin slept better than the man of ambition on a bed of eider-down.

Geese such as the Canada goose *Branta canadensis* also incubate their eggs in grass and down-lined nests. The most interesting is that of the barnacle goose *Branta leucopsis*, which breeds only in rugged, mountainous country in three widely separated areas – Spitzbergen, Novaya Zemlya and eastern

Greenland. The birds return faithfully to the same site each year, so that the cup gets larger as the grass and down, and the droppings of both birds, accumulate. They breed socially on wide ledges on rocky cliffs and pinnacles, situations more suited to raptors' eyries, which is no doubt an adaptation against raids from Arctic foxes. The population of this goose numbers only 50–60,000, which is relatively small by the standards of most birds, and everything must be done to protect its specialised wintering habitat in western Europe where it is all too vulnerable (Hudson 1975).

The first barnacle goose's nest was found in 1891 in Greenland, and that at last laid the mystery which till then had surrounded it. For centuries the flocks faithfully arrived each winter in Europe, and strange legends arose to explain their origin. Irish Roman Catholics used to eat the bird on Fridays and during Lent when they were forbidden to eat meat – they believed the geese hatched from black and white barnacles *Lepas* spp on the shore and so were really fish! The earliest record of this is in *The Journal of Friar Odoric*. He dictated an account of his Asiatic travels in 1330, and compared a strange discovery in the land of the Great Khan with 'as I myself have heard reported, that there stand certain trees upon the shore of the Irish Sea, bearing fruit like unto a gourd, which, at a certain time of the year do fall into the water, and become birds called bernacles, and this is most true' (see Pollard 1900).

Among the passerines, the pipits and many wagtails *Motacillidae* and the larks *Alaudidae* are particularly ground-loving birds. All the former build cups of grass, lined with hair and plant fibres, well hidden in hollows. The yellow wagtail *Motacilla flava* of Eurasia, for example, builds beneath the overhang of a clod of earth on fallow land, under the ample cover of cabbages or rhubarb in market gardens, or in a grassy tussock in water meadows. The hen alone does the work, taking as little as four days to complete the nest or as long as three weeks in bad weather (Smith 1950).

The larks, besides choosing sites with some cover, even in desert areas, have several genera containing species whose nests have, rather curiously, a rampart of small stones built up against the grassy cup on the exposed side. The horned lark *Eremophila alpestris* of the Holarctic provides an elementary ramp, but those of the finch larks *Eremopterix* spp of Africa are often much more obvious. They are variously described as being for decoration or for drainage after a flash flood. The bushlarks *Mirafra* spp, which range from west Africa to Australia, show an interesting step, or halfway stage, in the evolution of the domed nest; their cup-shaped nests are partly

This flappet lark *Mirafra rufocinnamomea* is one of the passerines which prefers to nest on the ground, building a partly-roofed cupshape of grass well-hidden in a hollow. (*J. F. Reynolds*)

roofed, some of the grasses of the tussock being bent over to form a thin cover.

Lastly, the world's largest passerine, the superb lyrebird *Menura novaehollandiae* of Australia, *usually* builds its bulky domed nest on or near the ground in thick scrub, but will occasionally nest well up in a tree fern; as much as 100ft ($30\frac{1}{2}$m) up in a substantial fork has been recorded. Built by the hen, the outside is made of sticks, with an inner wall of fine rootlets and bark. The cup is lined with the hen's own feathers from her back and flanks (Cayley 1950).

9 In holes

Many orders of birds contain species which nest in holes – in cliffs, in trees, in peaty turf, in the ground, in cavities under boulders. Some use natural holes, others excavate their own. At one extreme are the penguins, relatively primitive birds; the little penguin *Eudyptula minor* of Australia, for example, makes a nest of a little dry grass or seaweed in a cavity beneath rocks or turf, or sometimes in a petrel burrow. At the other extreme are the passerines, the world's most highly evolved birds, and many of them are hole-nesters too, such as the very successful starling *Sturnus vulgaris*. The 'old' families build little or nothing in the burrow or cavity, whereas the most highly evolved often build substantial, well-shaped nests as well as taking advantage of the safety of the hole. It seems clear that this is a result of convergence, which is evolution that produces an increasing similarity in a particular characteristic between groups that were originally different.

Why convergence has occurred in some families and not in others, and within those families not in every species, would make a stimulating study. Among the ducks for instance, the

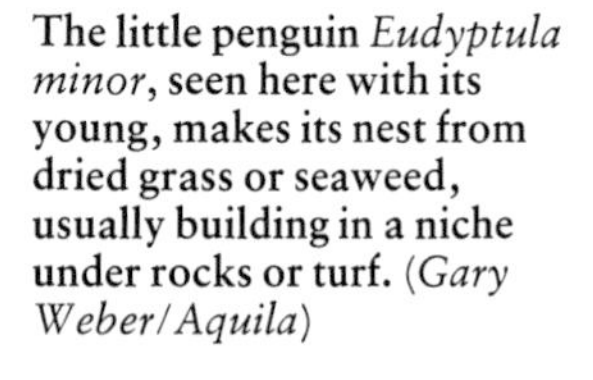

The little penguin *Eudyptula minor*, seen here with its young, makes its nest from dried grass or seaweed, usually building in a niche under rocks or turf. (*Gary Weber/Aquila*)

The shelduck *Tadorna tadorna* has a tendency to nest deep down a rabbit burrow. This one in the absence of a burrow has excavated a tunnel in a haystack. (*Brian Hawkes*)

shelduck *Tadorna tadorna*, which breeds intermittently across the Palaearctic, is well-known for its partiality for nesting 4–10ft (1–3m) down a rabbit burrow near the coast or an estuary. Of the 44 other ducks which breed in north America and Europe, only the ruddy shelduck *Tadorna ferruginea*, goldeneye *Bucephala clangula*, Barrow's goldeneye *B. islandica*, buffelhead *Charitonnetta albeola*, red-breasted merganser *Mergus serrator*, goosander *M. merganser*, smew *M. albellus*, hooded merganser *Lophodytes cucullatus*, black-bellied tree duck *Dendrocygna autumnalis* and wood duck *Aix sponsa*, habitually nest in holes. These divide into four groups – shelducks, goldeneyes, sawbills and 'treeducks'. Most ducks nest hidden in thick cover, but only these few have by natural selection become hole-nesters; indeed, goldeneyes in particular have taken to breeding in nest-boxes, and their firm establishment in Scotland is at least partly attributable to extensive erection of nest-boxes there (Sharrock 1976). Elsewhere in

Europe they are known to nest in the holes made by black woodpeckers *Dryocopus martius*, and in North America in those made by the pileated woodpecker *D. pileatus*. Nest holes have been reported over 13ft (around 4m) deep in the tree trunk, and others where the hole was 50ft (15½m) high. Yet from such nests the ducklings fledge successfully, dropping safely to the ground cushioned in down, and waddling in line astern of the duck to the nearest water.

Only one wader nests in a hole, the crab plover *Dromas ardeola*. It haunts the shores of the Red Sea and the Indian Ocean and is well named, for its main food *is* crabs. These birds nest among dunes, burrowing into the sand like petrels or puffins. Often, so many nest near to each other that it is difficult to walk across the nesting area without the ground caving in. The burrow, 4–5ft (approx. 1½m) long, leads downwards for some way, then curves up to the nest cavity, which is therefore above the level of the tunnel, possibly as protection against flooding. The crab plover has been nesting so long in the darkness of a tunnel that like most other hole-nesting birds

This Siberian tit *Parus cinctus* has taken over the old nest of a three-toed woodpecker.
(J. B. & S. Bottomley)

it has evolved a white egg, unlike other waders which nest in the open and, as we have already seen, have heavily marked, well-camouflaged eggs. So strong are this bird's unusual characteristics that it is placed in a family of its own.

The classification of birds is a very difficult task and systematists have long argued about the exact relationship of one bird with another, at one extreme, and even about how many orders there are, at the other. According to the Peters' system (see Gruson 1976 and Thomson 1964) these three orders immediately precede the *Passeriformes*: the *Trogoniformes*, *Coraciiformes*, and *Piciformes*. Orders are primary taxonomic categories and so contain species which are closely allied. Scientists who are sceptical about relationships tend to preserve more orders; those who are more confident place more families together, in fewer orders. It is interesting that all the 16 families in these three orders nest in holes. The parasitic honeyguides *Indicatoridae* are no real exception, because all but one of the host species are hole-nesters, such as barbets *Capitonidae*, which are members of the same order.

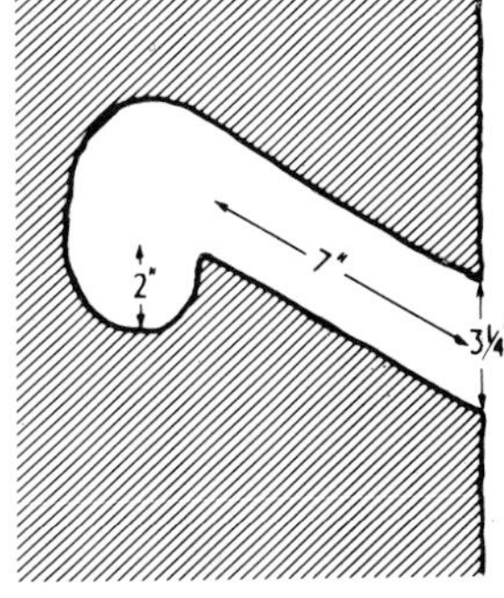

Above:
Longitudinal section of an enclosed chamber, carved in wood or in termitaries by the white-tailed and massena trogons *Trogon viridis* and *T. massena*. Loose particles of wood act as a nest-lining, but the nest is not cleaned and soon smells of ammonia.

The white-tailed trogon *Trogon viridis* ranges from Costa Rica to southern Brazil. The nest hole is excavated by the male and female working alternately. The bird *bites* at wood which is soft enough to dig one's nail in but not soft enough to crumble to dust in the fingers. If it is harder than that the birds cannot work; if it is softer, the egg chamber will not be strong enough.

Sharing of excavation work by male and female white-tailed trogons

Male worked for 5 minutes
Female worked for 9 minutes
Male worked for 2 minutes then frightened by falling branch.
2 minutes intermission
Male worked for 10 minutes
Female worked for 10 minutes then 2 minutes intermission
Male worked for 10 minutes
Female worked for 7 minutes.

As one bird dropped from the hole the other went in. Both worked rapidly. The diagrams show two basic types of trogon nest, enclosed and shallow (Skutch 1962).

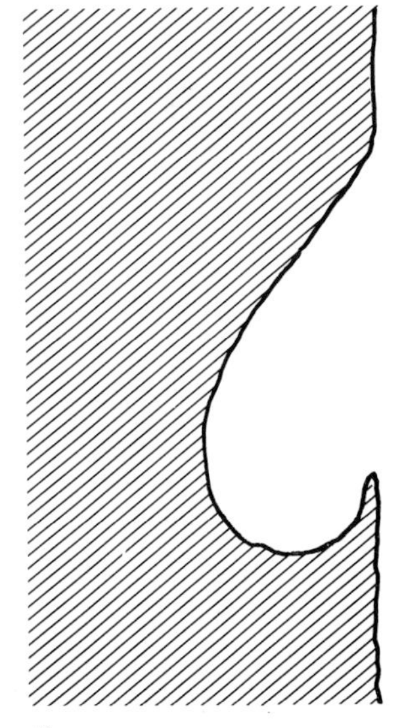

Above:
The shallow niche carved by the collared and Mexican trogons *T. collaris* and *T. mexicanus*, in decaying wood.

Kingfishers and bee-eaters are spectacular birds, many of them having exotic plumage as well as exciting behaviour. Both cock and hen tunnel into the soil or sand of a river bank. *Alcedo atthis*, the beautiful common kingfisher found in much of Europe, Africa and Asia, begins by flying repeatedly at the

The exotic, colourful white-throated bee-eater *Merops albicollis* digs its nest in sandy soil in flat country. (*J. F. Reynolds*)

bank, aiming a blow with its bill at a place well above the water level. Soon a depression is formed so that the bird can perch and then dig on more rapidly. The loosened soil is shovelled backwards with its feet until a rising tunnel up to a metre long is formed with a nest chamber at the end. The tunnel is only just wide enough for the bird and it cannot possibly turn round until it gets to the circular chamber at the end. There is no nest but the white eggs are placed in a litter of disgorged fish bones and pellets. The nest-chamber is usually clean but 'the tunnel is a running sewer of greenish liquid and decomposed fish and smells abominably' (Coward 1950). The larger kingfishers dig longer tunnels; that of the belted kingfisher *Ceryle alcyon* of North America may be 2m long, and the African lesser pied kingfisher's *Ceryle rudis* in soft soil or sand may be 3m. On average the latter takes four weeks to dig its home. There is some correlation between the feeding and nesting habits of kingfishers; the fish-eaters burrow in river banks, while the terrestrial feeders (such as the sacred kingfisher *Halcyon sancta* of Australasia) more often burrow in termite nests or use cavities in trees.

The blue-diademed motmot *Momotus momota*, a gorgeous blue and green bird of Central and South America, prefers to dig its nesting burrow from the side of a pit or hollow in the ground, such as the den of a burrowing animal. The mouth of the bird's tunnel may be invisible until the searcher sticks his

The malachite kingfisher *Alcedo cristata* builds in a river bank; note the shallow ledge filled with disgorged fish bones and pellets on the right. (*J. B. & S. Bottomley*)

head into the larger excavation. The digging begins long before the nest is needed. One begun in October did not house young till the following April. Another burrow found on 30 September was already 20½in (52cm) long and when finished on 5 November measured 80in (203cm). A third was 4in (10cm) on 1 September. Both birds shared the digging, doing about 2in (5cm) per day until on 9 November this one too was 80in long. The work had taken two-and-a-half months! The birds dig in the rainy season when the ground is soft, and rear their young in the more or less dry conditions of March or April. This is undoubtedly a successful system – predators appear to take less notice of the hole because the tunnel looks old. The motmots neglect it in the intervening months (Skutch 1964).

Among the most amazing nesting habits is the unique behaviour of the hornbills *Bucerotidae*. There are 45 species, all but two of them nesting in natural holes in trees. The habits of the red-billed hornbill *Tockus erythrorhynchus*, widespread in Africa in the scrub and semi-desert south of the Sahara, are typical of one group. Having found a suitable hole in a tree the hen collects mud and begins to plaster up the entrance. When the hole is almost too small she squeezes through and proceeds to seal it up, except for a small slit, using mud which has dropped inside and her own excreta. The male steadily feeds his mate with insects, fruit and small lizards, passing them through the small slit, calling perhaps 30 times a day through-

A colony of Layard's blackheaded weavers *Ploceus cucullatus nigriceps*. (*Eric Hosking*)

out the three weeks' incubation period. Two to five eggs are laid, hatching several days apart, and the male then has to bring food for the whole family, stepping up his visits to about 70 a day. To keep the home clean the hen ejects her excrement through the slit, and uses the brood's droppings to reinforce the plastering. After an imprisonment of 40 days she breaks out, a job which may take up to four hours, and joins in helping to feed the young.

Other hornbills do not escape but remain in the nest until the young fledge. The red-billed's behaviour seems to be an adaptation to make sure enough animal food for the young is collected in the increasingly dry season. As if the immurement of the female were not amazing enough, the young birds instinctively reseal the nest-hole after the hen has flown, using their own droppings and sticky berries brought by the cock – a remarkable achievement on the part of half-fledged nestlings. They take four days to do it. When they fledge, after six weeks, they break their own way out of the nest. The younger birds stay a few days longer than older siblings and begin to replaster the hole after the first bird leaves (Root 1969).

The silvery-cheeked hornbill *Bycanistes brevis*, of eastern Africa's coastal and montane forests, is typical of the species in which the female spends *all* the nesting period in the hole – about 45 days of incubation and 58 days for fledging (Moreau 1941b). The male brings all the food, mostly fruit, making about 16 trips a day. It is regurgitated one fruit at a time. Female hornbills undergo a complete moult when in the cavity. Their unique nesting behaviour creates a very safe sanctuary from predators such as tree snakes.

Jacamars' bills seem too delicate for digging a tunnel and it is hard to imagine they belong to the same order as the woodpeckers. The rufous-tailed jacamar *Galbula ruficauda* of central and northern South America is a forest bird. With its long needle-like bill it digs rather like a kingfisher into any steep earth or sand bank as little as a foot high, or even an earth wall thrown up by a fallen tree. The burrow may be used in successive years. Both male and female excavate, alternating in spells of from two to four minutes. Digging or renovating may take a week, but eggs may not be laid for another three weeks.

Nine jacamar burrows were studied in detail in Guatemala (Skutch 1963). All were within 11½–16½in (29–42cm) long, but the majority were 12–13in (30–33cm) in length. Their inside diameter was 1¾in (4½cm), some nests being a little wider than their height. Five tunnels inclined to the right of observers facing the entrance. The burrows expanded into a low vaulted chamber where the eggs were laid on bare earth with no lining.

The puffbirds *Bucconidae* are generally short-tailed and large-headed, their abundant lax plumage giving them a chubby appearance. One of the family, the white-whiskered soft-wing *Malacoptila panamensis*, inhabits primary rain forest from Ecuador to southern Mexico. It sits and waits on an exposed perch, appearing 'stupid', until a sudden dart for an insect, spider or even a lizard as big as itself proves the watcher wrong. Not many of its nests have been found. Curiously for a little bird (about 7in or 18cm long), which rarely hunts in bushy undergrowth, but whose ecological niche is the 'middle tree' from 15ft (4½m) up, it nests in a burrow in the ground. The entrance is surrounded by a low collar or frame of coarse dead leaf-stalks, axes of compound leaves and twigs, with leaves overlying them all. Whether this has been put there by the birds or has fallen there naturally is difficult to decide. The birds enter and leave by passing through the collar of sticks which blend with the leaf litter and help to conceal the entrance. The related black nunbird *Monasa niger* of the Orinoco basin may heap up a tunnel of half-a-bushel of twigs.

It is not clear whether soft-wings dig out their own nests; no soft excavated earth is outside the burrows studied, which suggests they take over existing holes, but the regularity of the burrow size and its directness weigh against this view. Probably puffbirds carry away any tell-tale soil (as tits *Parus* spp carry

Light-mantled sooty albatrosses *Diomedea palpebrata*. (*R. Gill/Aquila*)

The shore lark gathers stones for the foundation of its nest. This serves as both decoration and drainage. (*Dr C. J. F. Coombs*)

away wood-chips). So straight is the burrow that the brooding adult or the chicks are easily seen in torchlight. Incredibly, when the young stop being brooded by the male at night, a protective screen of dead leaves is erected at the entrance of the nest chamber. It is believed that the young raise this, from dusk on, by vibrating movements of their wings and feet. The screen may reach the ceiling, but its purpose is far from obvious. The light at night, even the brightest moonlight, does not reach down the burrow to the nest-chamber. Perhaps it is to fool bird-spiders and snakes: certainly soft-wing nests seem to be pillaged less often than the majority of woodland nests (Skutch 1958).

Woodpeckers have long attracted our attention by their amazing agility in climbing, and by their drumming on dead wood which resounds far through woodland. They excavate their own holes each year, the male usually doing more of the work. Although a cavity may take three weeks to hollow out, it is rarely used a second time by the woodpeckers, although, in later years, it may be taken by one of many species – stock doves *Columba oenas*, starlings, sparrows or tits, or even by

A male pied flycatcher brings grubs to his nest hole in a natural site. (*J. B. & S. Bottomley*)

squirrels and bats! Indeed, there are many records of other birds usurping the nest-hole as soon as the woodpeckers have finished all the excavating. Bullying starlings do not always win and there is one record of a pecking, clawing battle in which the great spotted woodpecker *Dendrocopus major* was victorious. The much larger black woodpecker is sometimes ousted by jackdaws *Corvus monedula*.

A woodpecker's bill, flattened laterally, is an excellent chisel. The rhythm of movement is quite different from that used when drumming. The bird deliberately swings its head, chiselling a wedge-shaped chip, taking from 3 to 15 blows for each chip. The hole often goes upwards at first so that rain does not drip in, but it soon turns down and becomes vertical. The small lesser spotted woodpecker *Dendrocopus minor* excavates only 9in (23cm) deep, but the black woodpecker, which is over three times as big, makes a cavity 2ft (61cm) deep. As soon as a cock great spotted woodpecker starts in earnest he 'is impelled by a feverish urge to work. He will hammer away up to six hours a day, often for more than an hour at a time', whereas cock and hen black woodpeckers usually change over every 40 or 50 minutes (Sielmann 1959). Towards the bottom of the cavity the diameter increases till the nest chamber is reached. The white eggs are laid on the wood floor, perhaps among a few chips, the rest of which are picked up by the beak load by the excavating bird and are dropped out of the hole periodically during excavation to litter the base of the tree.

Although most woodpeckers may appear to be digging into a live tree, careful investigation will show that the heart-wood is dead. In south-eastern North America, however, lives the

A pied flycatcher, with its hungry brood, seems quite at ease in its man-made nesting-box. (*E. V. Breeze Jones*)

red-cockaded woodpecker *Dendrocopus borealis*, which is unique in exploiting living pine trees. Nearly all its food is found under pine-bark; and the sapwood is often tapped so that sticky resin flows free. The birds deliberately keep resin flowing around the nesting hole. They have learnt how to keep their plumage clean, whereas visitors and prospective predators are deterred by the sticky tree-trunk which may be marked for many feet each side of the hole.

Another curious woodpecker *Geocolaptes olivaceus*, the ground woodpecker, is confined to the Union of South Africa. It has 'no interest in trees, but lives and gets its living on the ground' on barren, rocky hillsides (Gill 1936). Its eggs are laid in a chamber at the end of a tunnel which is usually a metre long and has one deliberately formed bend. The burrow is made in the bank of a flowing or dried-up river, or the earth bank of a road cutting.

A most unusual nest is that of the black wheatear *Oenanthe leucura*, a member of the thrush family with a very limited distribution in Iberia and north Africa. It breeds on rocky hillsides and cliffs, in a hole, crevice or cave in the rocks (or a building if it has a suitable hole in the wall). There has to be room for a foundation of stones, which are carried to the nest site by both male and female for up to two weeks before the actual nest is begun. Sometimes the stones are not that skilfully inserted and fall to the ground or slope below the hole, the sound carrying over 100 yards. The number of stones collected varies greatly from site to site; 358 have been counted in one nest, but the pile down below is often much greater. The accumulation of several years at one nest covered 2sq.yd of cave floor and sample counts gave an estimated number of 9,300. Their weights are often remarkable: the male wheatear weighs about 1¼oz (37½g), and when about 1,000 of the nest stones were weighed (Richardson 1965) the largest was as much as 1oz (30g). One record is of an almost unbelievable 2½oz (75g). Of 120 stones from one nest, 84% weighed between one-ninth and one-third of an ounce (3–10g). Stones are collected within 2–10m of the nest site, and are selected on the basis of size and shape, a flat one or one having a flat projection being essential if the bird is to hold it in his beak.

The gathering of stones for the foundation of a passerine nest is remarkable; it has been observed also in several desert larks, the rock wren *Salpinctes obsoletus*, eastern phoebe flycatcher *Sayornis phoebe*, and the white-crowned black wheatear *O. leucopyga*. The stones do not now seem to serve a protective purpose, although they may have helped deter ground predators in times past. Today they serve as a founda-

D'Arnaud's barbet *Trachyphonus darnaudii* **emerges from its nest – a hole in the ground.** (*J. F. Reynolds*)

tion for the nest proper, but the main purpose of the stone-carrying is to serve as an advertising display by the male in nest-site selection, which is continued as part of true nest-building. The black wheatear may have evolved into the largest and strongest wheatear in conjunction with this rock-carrying habit and its survival value. The black and white male must be very noticeable to a female and the sounds of falling stones probably provide auditory stimuli too!

The nest proper, set in the stony foundation, has an outside diameter of 8½in (21½cm) and a cup about 3in (7–8cm) across and 2½in (6–7cm) deep. It is made of grass, leaves and plant fibres, lined with finer plant material and several hundred feathers. From the start of rock-carrying it may take three weeks to build.

As we have seen, many other passerines nest in holes, but build a substantial cup-shaped or even domed nest within them. Among the flower peckers of Australia, the spotted pardalote *Pardalotus punctatus* builds a rounded, domed nest made of strips of bark at the end of a tunnel in an earth bank. The little tit-like birds make the 12–24in (say 30–60cm) tunnel themselves. The red-browed pardalote *P. rubicatus* also builds with bark, but makes only a cup-shaped nest. Many other small passerines throughout the world take over existing holes in trees or nestboxes.

Even the large stick nest of a species of crow can be made in a hole. In Europe the jackdaw normally builds in a cavity in rocks or trees, down chimneys, in lofts and thatched roofs, and even in rabbit burrows. There is generally a stick foundation lined with wool, hair, fur, paper, grass and other oddments. The nest may be slight, but when a site is used annually a mass of 2cu.yd (1½cu.m) may accumulate. This species' predilection for collecting bright objects and adding them to the nest has been immortalised in R. H. Barham's famous poem 'The Jackdaw of Rheims'. The bishop's ring was stolen and the repentant bird leads the way:

> Slower and slower
> He limped on before,
> Till they came to the back of the belfry-door,
> Where the first thing they saw,
> Midst the sticks and the straw,
> Was the ring in the nest of the little Jackdaw.

10 Aquatic nests

Of all the habitats in which birds live, water appears to offer the most impossible foundation for nests. Even the wildest, strongest-flying ocean birds return to cliff or rock to breed, and many so-called seabirds nest far from the sea – such as the mew gull *Larus canus* of the Palaearctic and western North America. Although finding sites near pools or streams, the mew gull may go well inland, to heather moorland or grassy islands on mountain lochs, as I have seen in Scotland. It can be found nesting in Norway 4,000ft (1,400m) above sea level in mountains (Voous 1960), and elsewhere even 30ft (10m) or more up in trees (Peterson 1961, Sharrock 1976). Even less aquatic in its breeding season is the grey gull *Larus modestus* of

The great crested grebe *Podiceps cristatus* either makes a floating nest or, if the water is shallow builds up from the water bed. The 'island' nest is usually tethered to some kind of waterside vegetation. (*Stephen Dalton/NHPA*)

the desert coasts of Chile and Peru, which nests up to 50 miles (83km) inland in a stony desert, in temperatures ranging from a night time 2°C to a midday 42°C!

Nevertheless some birds *have* adapted to an almost non-stop watery existence – grebes for example. The great crested grebe *Podiceps cristatus* is one of the most studied birds in the world, being the subject of pioneering behavioural studies by Selous and Huxley at the beginning of this century. It is widespread in Europe, Asia, Africa and Australasia on large lakes fringed with reeds, sedges and other aquatic plants. In its breeding plumage it is a beautiful bird and its spectacular courtship has entertained millions of viewers thanks to a Walt Disney film.

Grebes are 'efficient if not very elaborate nest-builders' (Simmons 1955). Most nests are tethered to waterside vegetation of some sort, with at least part of the structure awash so that the birds can swim right up to it: their legs are placed well back for maximum propulsion in the water, so that they are very clumsy on land and prefer to swim whenever they can. The nests may be well hidden, but some are in the open among lily-pads. Both sexes build, the male taking the larger share. Water weed, mostly sodden and decayed, is brought by the beakload; the birds dive for it around the nest site, sometimes going as far away as 50yd. In shallow water the nest may be built up from the bed of the pond, but otherwise it is simply a growing pile of rotten weed, with perhaps a few sticks in the base, anchored to some rushes or sedges. When actively building the male and female have been seen to bring 100 'cargoes' of

The little grebe *Podiceps ruficollis* adds green matter to its nest; possibly the fresh vegetation when added to decayed matter starts fermentation which could aid incubation. (*J. B. & S. Bottomley*)

material in 50 minutes. Most of the weed is deposited on the growing slope of the nest-platform, occasionally being pushed into place with a shovelling action of the bill. Later one bird will climb on the shapeless mass and pull at the weed to make more of a flat cone, then give the platform additional strength by deliberate heavy trampling with its large lobed toes – an action often accompanied by loud squelching noises!

There is no real cup but the incubating bird will rearrange a wall of weed around itself and, in bad weather, add material to raise the eggs above the water. Rankin (1947) saw one nest in bad weather when waves 'were frequently breaking against the upper edge of the nest, and every now and again a larger one would hurl itself against the breast of the sitting bird. To some extent the nest rose and fell with the water and one admired the skill of its builders in providing it with a certain amount of elasticity while at the same time keeping it firmly anchored'. Unhappily some nests do get flooded, or even left high and dry in drought. When an incubating grebe leaves its eggs, it covers them with weed, some birds performing this task better than others. Another curiosity of grebe behaviour is the way they carry their small young on their backs on fishing trips in the lake, using the nest-platform as a resting place on their return.

The horned or slavonian grebe *Podiceps auritus* and the eared or black-necked grebe *P. nigricollis* breed in North America and Eurasia on suitable areas of fresh water. These two are similar in size, smaller than the great crested. Although both build a 'soggy floating heap', *nigricollis* builds one in three hours, only 10–14in (25½–35½cm) across at water level, whereas *auritus* heaps up a large raft-like structure which may be more than 3ft across.

The little grebe *Podiceps ruficollis* has been observed adding *green* vegetation to its nest. The mixing of fresh with decayed material may set up fermentation, which might possibly assist incubation (Bird 1933).

Nearly all the members of the family *Rallidae* nest in marshes or by ponds and lakes. The coot *Fulica atra* is widespread in the eastern hemisphere (except Africa) and builds a strong compact structure usually hidden in reeds, yellow flag irises, rushes or reedmace. Made from surrounding vegetation, unlike a grebe's nest it does not float, but is built up from the bottom till it rises 6–12in (15–30cm) above the water. Some are 'so firm as to support the weight of a man seated when up to the knees in water' (Saunders 1899) although I know no one in more recent years who has tested one!

Coots are interesting for two other reasons. Firstly, the male is often enterprising enough to build a platform near the

The moorhen *Gallinula chloropus* is a close relative of the coot and builds a nest well hidden in reeds or rushes. The structure is built up from the bottom until it is clear of the water and is usually made of the surrounding vegetation. (*NHPA*)

The horned coot *Fulica cornuta* breeds on the mountain lakes of the Andes where there is a shortage of vegetation. The coot will build up a mound of stones in the shallows and then use aquatic plants for the nest material. (*Dr C. J. F. Coombs*)

nest where the young are brooded. Sage (1969) has recorded males building platforms long before the eggs hatched, but others leave it until the chicks are two days old. Some platforms have an approach ramp to enable the family to climb up; the chicks must be carried up to those without one. Secondly, less efficiently, coots are sometimes clumsy enough to knock eggs out of the nest (it has only a shallow cup). Occasionally, when they add extra material to the nest, they bury eggs so that they fail to hatch.

Far more curious is the nesting of the horned coot *Fulica cornuta* of the mountain lakes of the Andes. In Chile, in particular, in its arid mountain habitat there is little vegetation with which to build, and so the birds construct mounds of *stones* in shallow parts of the lakes (Johnson 1969). The mound is a huge cone, 2–3ft tall and up to 13ft (4m) in diameter. This brings it close to the surface of the water, and the coots build a nest of aquatic vegetation on top, so that the whole rises 1–2ft above water level. The vegetation is pulled up from below the surface. Although horned coots have been seen bringing stones one by one, weighing as much as a pound, from the shore to build up the mounds, it is still hard to believe that such constructions are their unaided work. Living as they do in such inhospitable territory it is likely that a certain mystery will remain for some while to come.

All seven jacanas of the tropics and subtropics are fascinating birds with attractively coloured plumage; they have spectacularly long toes and claws which enable them to walk with ease across floating vegetation – hence their other common names of lily trotters and lotus birds. Their nests have been described as 'almost an insult to the art of nidification'. The pheasant-tailed jacana *Hydrophasianus chirurgus* of India and south-east Asia condescends to make one which 'consists of a few floating water-weeds piled together to form a small blob, almost awash' (Phillips 1952). The lotus bird *Irediparra gallinacea* of Australia also builds a flat structure, on bunches of weeds or grass growing in the water. These and other jacanas' nests are so insubstantial that when the bird

The African jacana *Actophilornis africana* is a colourful bird with extraordinarily long toes and claws which enable it to walk on floating plants. Their nests, in which they lay beautifully patterned and glossy eggs, 'consist of a few floating water-weeds piled together to form a small blob'. (*J. F. Reynolds*)

stands up the nest and eggs are often thrust beneath the water. Indeed, sometimes the eggs are 'partly under water when the bird sits on its nest' (Cayley 1950). The eggs are distinctive: all have a very high gloss and only the pheasant-tailed's lacks a beautiful network of scrawls and twisting lines. The gloss is due to a waterproof cuticle which is essential to the survival of the species, for a normal egg would never survive the regular immersions these have to suffer.

One other group of birds is adept at living as watery an existence as the jacanas – the three marsh terns. The most widespread is the black tern *Chlidonias niger*, which is well distributed in the Holarctic wherever there are still fresh-water marshes or swampy grasslands with sheets of open water for feeding and extensive shoreline vegetation. On the Spanish marismas the nests 'can only be reached on horseback, or at least by thigh-deep wading; the birds preferring those stretches

The black tern *Chlidonias niger* builds a free-floating nest usually in an inaccessible part of Holarctic marshlands. *(J. B. & S. Bottomley)*

which will remain flooded until their broods are strong on the wing' (Blair 1962). The nests are usually found actually floating in shallow water. They are made of broken reed stems, water weeds and sedges lined with seed heads and finer stems. In open water they may be quite substantial, 2–6in (5–15cm) high and 6in (15cm) across. They may be in an area thick with water lilies or rushes, or in quite open water 1–4ft deep, the nest anchored by just one or two growing stems of an aquatic plant. In some places these terns nest in drier situations and then the nest is less bulky. Nests on the ground seem commoner in America than in Europe (Voous 1960).

The closely related whiskered tern C. *hybrida* which has a more tropical distribution in the Old World across to Australia, builds a larger nest than its relative, but it is much more flimsy. The reed stems and rushes are just heaped together with little or no cup and, because of the movement of the birds and the rise and fall of the water, the nest would fall to pieces were it not for constant rebuilding work by the pair.

The marsh terns appear to live in very similar habitats, and little is known about the precise niche each occupies – a detailed study would be fascinating and rewarding, for there are few birds as lovely to watch as these small fluttering 'butterflies' dipping lightly to the water's surface to feed, flying as lightly as a single feather.

11 Colonial nesting and nesting associates

Explorers, scientists, and holidaymakers have long been fascinated by the large, sometimes huge, breeding colonies of birds they find on their travels. To others, a single beautifully made nest may be even more of a wonder than the sight of thousands of birds nesting together.

Heronry, rookery, swannery, ternery and gullery are terms in common usage; loomery (of guillemots) and goonery (albatrosses on a Pacific island) less so. Antarctic scientists have established the use of 'rookery' for a penguin colony, and 'city' for a very large seabird colony is gaining ground. More spectacularly, a colony of seabirds with their tiny rocky island 38 miles offshore from Callao in Peru are known as 'Hormigas de Afuera', Ants Offshore.

For birds there are three possible kinds of social organisation for breeding (Van Tyne and Berger 1959).

1. *Colonial.* Many birds breed together but give each other no direct help in nest building or rearing the young. Such species include auks, gulls, terns, herons, swifts, oilbirds, swallows and weavers; also some crows, babblers, thrushes, woodswallows, starlings and icterids.
2. *Co-operative.* A few species live on intimate terms in a close community, assisting each other in building one huge nest in which each pair has a private compartment.
3. *Communal.* The most highly organised bird communities may be found among a few passerines, in which several adults build a nest, two or more females lay eggs, and all the adults share incubation and feed the nestlings.

What is the benefit of colonial nesting? What is the influence of the mutual stimulation? Why are seabirds, in particular, so commonly colonial? What are the smallest and largest numbers of birds needed for a viable colony? Much more research will be needed before the final answers to these and many other queries may be found. Plenty of theories have been put forward, but there is little agreement among the authorities on bird behaviour. Some scientists, led by Fraser Darling, believe the social displays of colonial birds not only lead to synchronisation of the sexual cycle of the male and female, but also stimulate other pairs. Edward Armstrong believes that the

The shag *Phalacrocorax aristotelis*, which has evolved its dark plumage for social inconspicuousness, often shares its breeding ground with the kittiwake, which has evolved a light-coloured plumage for the opposite reason. (*Arthur Gilpin/NHPA*)

assembly merely serves to ensure that the male and female meet at the appropriate time facilitating sex recognition. There seems to be no simple all-embracing explanation; for example, it has been shown that seabird plumage types have evolved in a way that promotes social conspicuousness in the light-coloured species, such as gulls and gannets, whereas selection for social inconspicuousness has been a major factor in the evolution of dark-plumaged seabirds, such as cormorants (Simmons 1972).

Birds which prefer to live in loose-knit groups, such as the Old World linnet *Carduelis cannabina*, or the American goldfinch *Spinus tristis*, may be called 'colonial', but generally the term is applied to those listed at the beginning of this chapter. It has been argued that *all* birds are gregarious and that 'one of the important functions of territory in breeding birds is the provision of *periphery* – "periphery" being defined as that kind of edge where there is another bird of the same species occupying a territory' (Fraser Darling 1952). James Fisher (1952) following his study of fulmars *Fulmarus glacialis*, was inclined to doubt the 'Threshold Theory' and suggested the 'Young Pioneer Theory' – that the lower breeding efficiency of smaller colonies is due not to small numbers, but to the higher percentage of inexperienced young birds. They arrive late at a colony, breed inefficiently or not at all, and depart early, so effectively they are driven from the place where they hatched and search for new breeding grounds.

Some believe that among seabirds colonial nesting has resulted from the limited supply of nest sites, several species inhabiting the same cliff but surviving successfully because each has different food requirements.

Also, birds may nest colonially because a colony is a successful anti-predator mechanism – there is safety in numbers, with more eyes to spot the piratical skua or fox. Harrison (1973) has described how, in the Arctic, predator-prey relationships are far more complex than that. There is little time in the short summer for birds to have second clutches and predation by skuas and gulls must be serious. In 1963 he visited a tiny island in Baffin Bay, and found 414 eider nests, with 20 pairs of glaucous gulls *Larus hyperboreus* and two pairs of great black-backed gulls *Larus marinus* dotted among them! Records from 50 years before show there had been 300 eider nests and no gulls. One eider duckling had just hatched in a glaucous gull's nest, which was curious to say the least. A real explanation of the 25 per cent increase in eiders in association with the establishment of a colony of big predators would be fascinating. It appears that the eiders are being protected by the very birds which elsewhere are their predators.

In the original hardwood forests of central North America lived the large, gracefully proportioned passenger pigeon *Ectopistes migratorius*. In 1899 the last wild one was taken in Wisconsin, and the last individual died in 1914 in Cincinnati Zoo, Ohio. That it should have become extinct in historical time is the more remarkable because great flocks of the migrating pigeons used to darken the sky, and the weight of their numbers broke great branches from trees.

The simple nest of sticks and twigs was often piled so carelessly on a branch that the whole thing would fall in a breeze. Each pair usually reared only one chick, breeding in small colonies scattered throughout the range of the species. But for reasons not really understood the natural population-regulation mechanisms sometimes did not work. Then in a year of plenty the flocks would again gather in the sky, the most impressive gatherings of birds ever seen by modern man. After carefully sifting the old records, Greenway (1967) believes that Alexander Wilson's contemporary account in his monumental *American Ornithology* (1808–1814) is not inaccurate. Wilson estimated that one flock in about 1810 contained a staggering 2,230,272,000 birds! One area of 850sq.miles in Wisconsin used for nesting probably had 136,000,000 birds in 1871, with up to 100 nests in one tree. These very numbers were too great a temptation for hunters with guns. Organised shoots would blaze away and kill thousands and thousands in a day. So the passenger pigeon's unique gregarious disposition was its downfall. No more will the huge flocks arrive in spring in the forests of the north from their wintering grounds in the Gulf of Mexico.

Today the world's largest colonies are those of the little auk or dovekie *Plautus alle*. It is restricted to the Arctic, breeding principally in Greenland but also on Spitzbergen and the large islands off Siberia. In the Thule district of Greenland it breeds in countless numbers, the total population reaching about 30,000,000; and large colonies totalling 10,000,000 birds are found on the east coast in the Scoresby Sound area (Freuchen and Salomonsen 1959).

The breeding places are usually in gigantic screes which cover the slopes of the sea cliffs, the birds laying their single eggs on the ground with no nest, under stones or in crevices. The little black and white auks may be seen dotting the slopes 'like pepper and salt', calling continuously their characteristic purring 'piuli, piuli, piuli'. This is the Eskimos' word for the call, and it was the name they gave their friend and benefactor the explorer Robert E. Peary. So many birds cover the ground that when they fly off together one can hardly understand how they are able to avoid crashing into each other. The Eskimos harvest both birds and eggs, apparently with no effect on the population. Salomonsen has described how the birds are caught, mostly by the women, in nets and stored 700–800 at a time in a *giviak*, a seal-skin bag. During the winter such a *giviak*-full is proudly produced as a great delicacy for parties or big celebrations. The large eggs of the little auks are also collected, by the thousand, and are thought to possess a strange power.

The guillemot or murre *Uria aalge* is well known on both sides of the Atlantic; the birds crowd narrow seacliff ledges or the tops of stacks. The single egg is laid on the bare rock, with nothing beneath it, often only inches from a precipitous drop into the sea. It is incubated by both sexes for 28–35 days, resting on the webs of the birds' feet, covered by the belly feathers.

Right:
The guillemot ***Uria aalge*** **nests on densely packed, narrow ledges and lays a pyriform egg of a beautiful colour and design. Each egg is different, possibly to help the parent birds identify their own.** (*Richard Vaughan/ Ardea*)

Far right:
A Brunnich's guillemot ***Uria lomvia*** **nests in close association with common guillemots in north Norway.** (*Richard Vaughan/Ardea*)

The young are tended by both parents for about three weeks. Only partly grown they then tumble into the sea, no matter what the height of the nest site. They cannot fly for another three weeks or so, and are probably fed for a time, some authorities say by an adult which is not necessarily their parent.

A guillemot's egg is very interesting: its volume is as great as a golden eagle's, it is a beautiful pyriform shape, showing great individual differences in ground colour and markings. The colour may vary from white through shades of buff, brown and cream to blue or green. There is even a red kind. The brown or black markings may form spots and blotches or intricate scribbles, but some eggs have no marks at all. Guillemots are densely packed on the ledges and it is thought that the eggs' variations enable birds to recognise their own (Harrison 1975). The conical shape may have evolved to help the egg stay on the ledge; if moved it should roll *around*, the pointed end acting as an axis, but 'the idea that its shape allowed it to twist round in the wind is exploded, as is the egg when a careless sitter sends it hurtling to the rocks beneath; the frequent struggles for the rights of a ledge lead to many catastrophes' (Coward 1950).

At the other end of the number scale might be placed the waved albatross *Diomedea irrorata*, all 3,000 pairs of which breed on tiny Hood Island in the Galapagos. Although so rare, it is apparently maintaining its numbers, although the odds are against this – the already small population, the problem of navigating the Pacific ocean every other year to find the speck of lava rock where it must breed, and the difficulties of this strange nest site. Landing and taking off are difficult for an

One pair of the 3,000 waved albatrosses *Diomedea irrorata* all of which nest on the minute Hood Island in the Galapagos. The albatross, which lays a single egg weighing nearly half a pound, is maintaining its numbers despite the perilous difficulties of its nesting-site. (*Adrian Warren/Ardea*)

albatross with an 8ft wing-span and Hood Island has no fine natural runways – the ground is thickly strewn with boulders and spiny bushes, resulting in many bad landings and clumsy take-offs. Leg and foot injuries are common. One would expect the birds to breed by the cliff edge so that they could quickly become airborne, but oddly they nest up to hundreds of yards inland, penetrating dense scrub, so giving themselves enormous problems on every journey. After an intensive study Nelson (1968) could find no reason why they chose such nest sites; nor why the population stays so small (but note that Harris (1974) records that there are now about 12,000 pairs) when there seems to be plenty of spare (and we would say better) nesting space, presumably an abundance of food, no albatross competitors and no known natural predators.

The fascinating life of the waved albatross is worth reading in detail; after a very elaborate courtship, which includes a complex ritualised dance, the single egg is laid on bare ground, with not even a pretence at nest building, real or symbolic – very different from the behaviour of the wandering albatross (p. 90). Sometimes the egg, which weighs nearly ½lb (225g), is laid in a crack between boulders and cannot be incubated, and the hen seems to be 'too dull to roll it out with her beak'.

Travelling southwards about a thousand miles we come to numerous islands on the continental shelf of Peru and Chile. Here is one of the world's most extraordinary sights for bird-watchers. In arid conditions and safe from predators, three species in particular are breeding in millions: the Peruvian cormorant *Phalacrocorax bougainvillei*, the Peruvian booby *Sula variegata* and the brown pelican *Pelecanus occidentalis* – the *guanays*, *piqueros* and *alcatraz*.

The geographical and climatic factors have resulted in conditions ideal for the deposition of *guano*, vast quantities of the nitrogenous dung of these fish-eating seabirds. In the dry climate the birds' droppings are not washed away, nor is much of the nitrogen compound and potash washed out from them. Long ago the Incas of Peru knew the value of guano as a fine fertiliser, and the Spanish colonisers used it extensively. But not until the nineteenth century was the resource exploited; once the commercial potential was realised guano became an industry, and by 1875 about 20,000,000 tons had been exported to Europe and the United States. Coastal islands were levelled to bedrock; 2,500 years of accretion disappeared in less than 100 years. The authorities realised that the harvest had to be controlled, just as the Incas had centuries before. Islands are now dug every other year, all the principal bird colonies are reserves and are wardened by *guardianes*, and on the mainland 'islands' have been created by building predator-proof walls behind which birds have started to nest.

The guano collected today is principally the birds' nests – which are just a substantial rim of excreta and a litter of moulted feathers. On the Guañape Islands, for instance, it is impossible to find an inch of ordinary rock or earth. There is only the yellowish suffocating guano. In the height of the season, the millions of birds and the glittering 'icing' cap make a bewildering sight. The mind boggles at such huge counts of birds, but a serious estimate of 5,000,000 *guanays* for Central Chincha Island alone was made in 1926, and Las Viejas Island's chief guardian told R. C. Murphy, the great expert on South American seabirds, that there were 8,000,000 birds there. Murphy (1959) wrote in answer to that: 'I had no intention of arguing with that tough old customer. There probably were no more than a million birds on Las Viejas, but they looked like a billion'.

Such vast numbers need a staggeringly large quantity of food, and thanks to the cold upwelling Humboldt current there is a lush pasturage of plankton to support the fish *anchovetas* on which the birds feed. But every few years a deadly warm current called *El Niño* – the Christchild, because it often comes around Christmas – surges southwards, heating the water, creating havoc. The marine life perishes, hydrogen sulphide bubbles out of the sea, guano birds are poisoned, and breeding is minimal. Enormous catastrophes are relatively rare, being recorded for 1891, 1911–12, 1925–6, 1932, 1939, 1957–8, and 1965. Nelson (1968) records that the Peruvian 1957 population went from 26,000,000 to 6,500,000, and in 1965 from 17,000,000 to 500,000. However, many birds were away at sea and later in 1966, 3,000,000 were counted on the islands. The colonies are now also threatened by the anchovy-fishing industry. Peru has in a few years become the world's leading producer of fishmeal – for fertiliser. But the Peruvian government has strictly limited the catch, and carefully controls the guano too, trying to avoid killing 'the goose that lays the golden egg'. The birds' trump card is that they produce a better and vastly cheaper product!

Many people have sat in summer on a cliff top and enjoyed the therapy of watching a colony of gulls, such as the dainty kittiwake *Rissa tridactyla*. The world of the much-maligned herring gull *Larus argentatus* is also full of non-stop activity, with a complicated family life that would well repay your looking at it more carefully. Somewhat more exotic is the rare Heerman's gull *Larus heermanni*, which breeds only on islands off the coasts of north-west Mexico. Strangely, it is a dark-grey

A herring gull *Larus argentatus*, with its chick, has made its nest on a roof top in St Ives, Cornwall. (*J. B. & S. Bottomley*)

gull with a black tail, nesting in a desert-like climate. On Rasa Island in the Gulf of California only 2in (5cm) of rain fall per year, and the gulls sit on their eggs to keep them cool rather than warm! Boswall (1976) found that on the same island were 55,000 gulls and dense colonies of royal terns *Sterna maxima* and elegant terns *S. elegans*. Half-a-million seabird eggs a year used to be collected for food, but the colonies have been protected since 1964. Once or twice a day the thousands of bickering birds uncannily fall silent, and may be flushed, settling for safety on the sea – when a peregrine *Falco peregrinus* comes hunting. This led the birdwatcher there to say 'We see the falcon that killed the gull that ate the lizard that caught the fly that fed off the fish caught by the skill of the osprey'.

Among terns, the black noddy *Anous tenuirostris* is unusual in being a cliff or tree nester. As with the kittiwake, another cliff-nesting species which has diverged from ground-nesting stock, a number of the black noddy's peculiarities can be related to its nesting site. It has lost its cryptic nestling and juvenile plumage, and the habit of dispersing eggshell and droppings; and has acquired more elaborate behaviour patterns

A nesting colony of crested tern *Sterna bergii* in Australian pigface creeper shows in what close confinement the breeding birds live. (*Gary Weber*)

associated with the nest. About 75,000 black noddies are residents on Ascension Island in the Atlantic, breeding on seaward-facing cliffs (Cullen and Ashmole 1963). Apart from guano, feathers are the only nest-material used in significant quantities here, though elsewhere rubbish from the sea, seaweed, leaves, sticks, and grass are worked in. On Ascension, every regular noddy perch soon acquires a covering of droppings (guano stalactites) and ledges become wide enough for nesting, especially where they are protected from the sparse rainfall. The black noddy uses the guano to enlarge the ledge and bind its nest, as the kittiwake uses mud. Spasmodic building may begin as long as seven weeks before egg-laying. Typically the male collects any extra material and the female forms the nest by pressing the material into shape with her breast and trampling with her feet.

Also on Ascension are up to a million sooty terns *Sterna fuscata*, which rather oddly breed every 9½ months (Ashmole 1963).

Away from the world of seabirds, the magpie goose *Anseranas semipalmata*, of Australia, is colonial. It is confined to swamps and breeds in the rainy season, from January to June. At a colony of 293 nest sites near Darwin, the birds began building activity two months before eggs were laid. Many males had two females. The geese moved about the swamp, building several reed-platforms, 'stages' on which they preened and courted. These stages were made from growing rushes which were bent flat and trampled on. Nests proper were built in 10–36in ($25\frac{1}{2}$–$91\frac{1}{2}$cm) of water, amongst rushes up to 4ft tall, by adding uprooted rushes to a platform. More and more rushes were added each day until the eggs lay in a deep cup. A path through the reeds was made where vegetation had been gathered (Davies 1962).

All bee-eaters nest in chambers at the end of quite long tunnels excavated in sandy or soft-soil banks. Some nest singly, but several species are colonial, none more so than the carmine bee-eater *Merops nubicus*, found south of the Sahara. Its colonies of up to a thousand pairs are a breathtaking sight, which fully compensates for the strong ammoniacal smell from the nest burrows.

Unlike most other members of the swallow family, the sand martin *Riparia riparia* does not build a mud nest. It is a colonial bird, found on both sides of the Atlantic, breeding in sand pits and river banks. Despite their apparently weak bills and feet the martins excavate straight tunnels in the sand, 3–4ft long. The nest chamber at the end is carelessly lined with material which has probably been collected in flight – grass, straws, feathers. The colony may return year after year to the same holes if they have not been eroded. Sometimes unusual sites are favoured. I know a colony in a sea-cliff of ancient Devonian sandstone, another in the side of a flooded china-clay pit and a third in clay drainage pipes in a wall which forms one bank of a stream by a large dairy.

In the mountains of north-western South America live the oilbirds *Steatornis caripensis*. They are distant relatives of

The magpie goose *Anseranas semipalmata* first builds a reed platform, a stage on which it will preen and court. In order to make the nest proper the bird adds uprooted rushes to the platform until it is well above water. (*Dr C. J. F. Coombs*)

nightjars, but unlike their insect-eating kin they feed on the fruits of palms and laurels. This diet gives the young birds an exceptionally high body-fat content, so local Indians and farmers raid the nest sites for squabs, which they boil to render into fat for cooking. The oilbirds nest in great numbers on ledges deep in mountain caves, and when alarmed their deafening calls sound like all the foxes in the world chasing all the hens (Ross 1965). The local name is *guacharo*, Spanish for 'one who cries and laments'. Amazingly these large birds (they have a 3ft wing-span) bounce echoing click-calls off the cave walls in order to fly in the darkness – as bats do. They eat only the outer flesh of the fruit they collect, and the dropped seeds are used to form the cup-like nests – and also form a thick bed of detritus on the cave floor which heaves with scavenging insects, including one of the world's largest cockroaches 4in (10cm) long and whip-scorpions with their 18in (45¾cm) span of antennae-like front legs.

We have already described in detail (p. 72) the quelea's nest, but not its colonial behaviour. In east and south Africa it is one of the most serious pests of grain crops; though it weighs only ½oz (14g) it has been calculated that a million birds could destroy 60 tons of grain a day. The quelea has been studied in great detail at the request of several governments who wish to control it, and who have burned, bombed and poisoned colonies to try to save crops. A colony by Lake Chad, in acacia woodland, was one of a series which extended for 50 miles. It was estimated to contain between 500,000 and 5,000,000 nests – and its nearest neighbour four miles away was even larger! Vesey-Fitzgerald (1958) studied more carefully a colony in Rukwa Valley, in south-west Tanzania, in a stand of pure *Sorghum macrochaeta*, a robust grass 10ft (over 3m) tall. The colony was T-shaped, its stem being 1,000yd (915m) long, covered 93 acres (37½ hectares), had five nests per square metre in a sample area, and an estimated total of 2,500,000 nests. The birds were said to have arrived on 10 April, and a week later nest building was finished! It is no wonder that this little seed-eater causes such concern when it arrives like a plague. The numbers were evidently exceeded by the passenger pigeon, but are otherwise without parallel among land birds. It is also of interest in its remarkable convergent resemblance to a North American icterid, the redwing *Agelaius tricolor*, whose colonies have numbered 200,000 pairs (Lack 1966).

Compared with the colonial birds, very few species nest co-operatively. The quaker parrot *Myiopsitta monachus* of South America is unusual in a family of some 315 species for *building* a nest (most of the others use natural cavities, sometimes lined with leaves or bark). The nest is of twigs and is built in a tree. Single nests may be made but often a number are built contiguously to form a single great mass, with individual entrances leading to separate nest chambers. Twigs up to 20in (50cm) long are carried in the bill by both sexes. First a downward-slanting sheaf is formed, then twigs are pushed in to form walls and a roof. Projecting twigs are tucked in or chewed away, the shredded remains forming a lining to the nest hollow (Harrison 1973). A dozen or more pairs may co-operate in the building. Quite often a Brazilian teal *Amazonetta brasiliensis* has been found nesting quite happily in one of the apartments.

Probably the most remarkable weaver's nest is that of the famous sociable weaver *Philetairus socius* of south-west Africa. It is as social as any bird could be: it travels in flocks, feeds in flocks and breeds in large compound nests. The nests vary in size according to the size of the flock. Basically they are large straw cones with entrances in the base. First the whole flock constructs a roof of coarse dry grasses in a large tree, and under this a number of nest chambers are made, by nipping off the straws to form a tunnel upwards with a chamber at the top. Each pair of birds has its own nest-chamber (Roberts 1940).

The sociable weaver *Philetairus socius* of south-west Africa builds an enormous compound nest which the whole flock shares, each pair having its own nest-chamber. (*G. J. Broekhuysen/Ardea*)

A sociable weavers' nest built of dried grasses and straw, hangs from the branch of a tree. (*Clem Haagner/ Ardea*)

A small nest may measure 3ft (1m) in diameter at the base, be 3ft high and have ten entrances; a large nest for about 95 pairs had a base 25ft ($7\frac{1}{2}$m) long and 15ft ($4\frac{1}{2}$m) wide, and was about 5ft ($1\frac{1}{2}$m) high. Nests are added to and repaired throughout the year, and sometimes become so large they break the branches; they are very obvious and easily counted. Friedmann (1949) found 26 in an area 100 miles long by 10 miles wide (160 × 16 km), which he believed were practically all that existed.

Unlike many species of passerines in which each pair defends a territory in the breeding season, the whole flock of sociable weavers has a territory. Many other weavers, such as the black-headed or village weaver *Ploceus cucullatus* are colonial, as are house sparrows *Passer domesticus;* these will build their untidy straw nests full of feathers in the tops of pines and firs, and even in country hedgerows, as well as under your roof.

The palm chat *Dulus dominicus* of Hispaniola builds a communal nest too, a construction the size of a bushel-basket, up to 3ft in diameter, placed conspicuously at the base of the fronds of a palm tree. Up to 30 pairs may build the 'apartment

The sociable weaver will even use telegraph poles as the support for their spectacular structures.
(*Anthony Bannister/NHPA*)

block' together, each pair having its own separate compartment and private entrance from the outside. The nest chamber is lined with soft bark and grass. The birds roost in the communal nest outside the breeding season. It is difficult to say whether palm chats have evolved such a strange nesting behaviour in their island isolation, or whether they maintain the characteristics of species long extinct (Austin 1961).

Finally, in the social organisation pyramid, come those species which nest together *and* help each other in rearing their broods. The best-known are the cuckoos of South America, anis *Crotophaga* spp. Flocks of 6 to 12 birds live in bushy country, feeding on animal ticks. A clumsy twig nest is built in a tree, begun by one female who is then joined by the whole flock. One or more females will lay in the nest, and when the young hatch everyone helps to feed them. There is remarkable harmony in all the community's dealings with one another (Armstrong 1947). The guira cuckoo *Guira guira* behaves similarly save that its nest is, by all accounts, even more primitive and sometimes so thinly constructed that eggs fall out.

The greycrowned babbler *Pomatostomus temporalis* of eastern Australia is popularly known as Happy Family, as the 8 to 12 birds of the flock co-operate in building the nests, sometimes half-a-dozen nests, although apparently one female lays in each.

Non-breeding brown jays *Psilorhinus morio* have been recorded as helping build a nest for an adult pair, and later feeding the incubating female (Armstrong 1947), and some communal behaviour has been recorded in several other Central American jays. It may be a developing characteristic, for the Mexican jay *Aphelocoma ultramarina* is known to help another family, but a major diversion in that community is the habit of stealing nest material from nests already complete!

The ostrich harem is a simple form of communal nesting behaviour. Most commonly a male has three females, all of whom lay in the one nest, but incubation is usually performed by the male and a 'favourite' partner. Recently, Grimes (1976) has summarised all that is known and suspected about communal nesting in Africa. Rather surprisingly he lists 52 species which could fall into the category. For example the ground hornbill *Bucorvus leadbeateri* lives in groups which may consist of a mated pair, several other adults of both sexes and one or two juveniles. The adult males help to line the nest hole in a tree, and all the adults share in feeding the incubating female and chicks (which are not walled in). The yellow-billed shrike *Corvinella corvina* is gregarious, often living in flocks of a dozen or more. A group shares one nest with eggs laid by one female, but all the birds feed the nestlings. The same female has been recorded as the group's breeding bird for three successive seasons.

Up to four extra adult helpers have been observed feeding nestling arrow-marked babblers *Turdoides jardineii*. In another commune, believed to consist of three adults and three immatures, all the adults built one nest. After that had been robbed, they and a fourth adult built a second nest. Perhaps most strange of all, a pair of babblers with two or three helpers helped to rear a cuckoo. Grimes believes careful research will reveal much more evidence of communal nesting and will enable a clearer comparison with conventional nesting.

Apart from these birds that nest with their kin, there are many examples of 'nesting associates', birds that nest with species other than their own. Three main types of association have been observed.

1. *With other birds*. Herons, ibises and anhingas may be found all in the same colony. One or two turnstones *Arenaria interpres* not infrequently nest among gulls. Small birds such as

Riparia riparia **the sand martin, as its name implies, excavates a burrow in sandy soil.** (*Dr K. J. Carlson ARPS*)

grackles, wrens or sparrows have been found nesting in the massive bulk of an osprey's nest. Putting the last example somewhat in reverse, pygmy falcons *Polihierax semitorquatus* almost always nest in a chamber of a sociable weavers' nest, or in the thorny nest of the buffalo weavers *Bubalornis* spp. They never molest the weavers, who are not known to mob the bird of prey.

2. *With vertebrates.* We have already noted how many birds have adapted to nesting on man's buildings. The only other known mammal associate is the vizcacha *Lagostromus trichodactylus* of Argentina, in whose burrow on the pampa nests the common miner *Geositta cunicularia*, a member of the same family as the ovenbirds. In New Zealand, the shearwaters *Puffinus* spp are regularly found breeding in the occupied burrows of tuatara lizards *Sphenodon punctatum*.

3. *With insects.* This is probably the most interesting and most studied association. Several Asian and African species such as kingfishers and woodpeckers habitually dig holes in termites' nests, as do trogons and jacamars in South America, and the beautiful, white-tailed kingfisher *Tanysiptera sylvia* of northern Australia and New Guinea. As these birds dig into the termit-

12 Edible nests

In 1678 John Ray, the author of the first bird book written in English recorded: 'In the Sea-coast of the Kingdom of China, a sort of small parti-coloured birds of the shape of Swallows, from the foam or froth of the sea-water dashing against the rocks gather a certain clammy glutinous matter, perchance the Sperm of Whales or other fishes, of which they build their nests. These nests are esteemed by gluttons great delicacies, who dissolving them in Chicken or Mutton broth, prefer them far from Oysters, Mushrooms or other dainty morsels.'

The fantasy which this account contains – that the nests are made from fishy substances – is no more amazing than the truth. Since at least the ninth and tenth centuries the Chinese have indeed esteemed birds' nest soup as a delicacy, often eaten with meat dishes. Though modern scientists have shown that the nests have almost no food value, gourmets insist on eating them – as a 'delicious and nourishing consommé' – and so harvesting is big business in parts of Java, Borneo and the Philippines.

The birds which produce edible nests are the swiftlets *Collocalia fuciphaga*, *Collocalia vestita* and *Collocalia maxima*, all of south-east Asia. Many nest in sandstone caves along the coast; however, the most studied colony is that of the Great Niah Caves in Sarawak (3° 47′N 113° 48′E), where three species breed, *Collocalia maxima*, *Collocalia salangana* and *Collocalia esculenta* (but only *maxima* produces edible nests).

The nests are deep in the system of caves. The Niah cave is very irregular, divided into many chambers, tunnels and grottoes: one cavern could hold St Paul's Cathedral and have room to spare. On the walls and roofs are glued the thousands of nests of up to 1,500,000 adult swiftlets, which build from the entrance, through the twilight zone, to the black interiors – plus about 4,000,000 bats!

The swiftlets *fuciphaga* build 'white nests' and *maxima* build 'black nests'. Both are constructed from the birds' saliva, but 'black nests' include impurities of feathers and sometimes vegetable material which have to be removed before soup can be made. These nests are therefore much less valuable.

The birds' saliva glands are much enlarged in the breeding

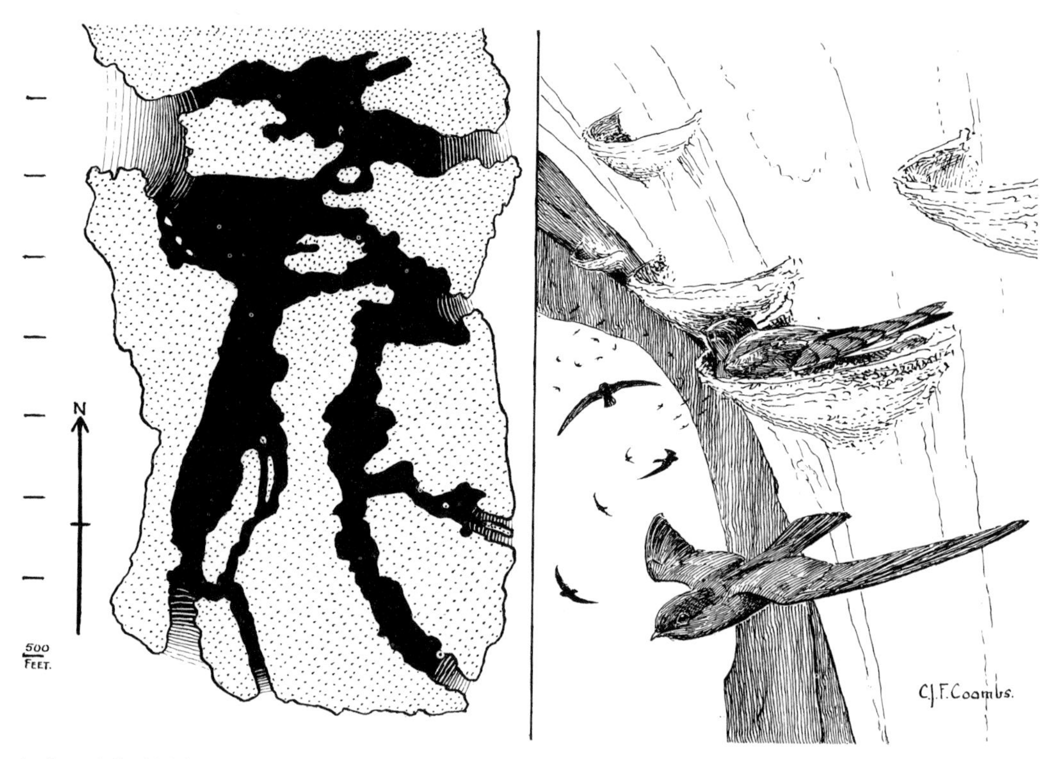

A plan of the highly complex system of Great Niah Caves in Sarawak, which houses the breeding colony of *Collocalia maxima*, one of the swiftlets which produce an edible nest. (*Dr C. J. F. Coombs*)
Collocalia maxima breeds in close groups, each nest being high up under the roof of the cave. The nests are made of the swiftlets' saliva plastered on to the cave walls in layers. (*Dr C. J. F. Coombs*)

season, producing quantities of glutinous, nearly white saliva. The nesting swiftlet builds by flying at the cave wall, just touching it with its tongue to deposit a drop, which soon hardens. First a horseshoe-shaped foundation is formed, then the base is built up to give a shallow cupped nest. The birds build principally at night, adding layer upon layer of 'nest cement', so that the nest is clearly laminated. No swiftlet has been seen collecting feathers so it is inferred that they use their own. There is little opportunity for the swiftlets to find nest support on the cave wall; the saliva ensures the nest is self-supporting. Usually the nests are from 8½ft (2½m) above the ground – which is feet thick in guano and crawling with invertebrates. The maximum height to which they build is limited only by the height of the cavern, which may be over 300ft (91m). The lowest found anywhere was at 6ft (2m) (Medway 1962).

New nests are pliable. At first 'nest cement' is translucent white and soft, but it hardens with age and deteriorates after a few months, especially in damp caves, going opaque yellow-brown, soft and friable. In dry caves nests persist through several seasons. Nest fall is hastened by the activity of the moth *Pyralis pictalis*.

In general dark sites are preferred. The swiftlets are able to fly safely in the caves by echo-location, like bats do. At Niah a huge population of *maxima* has resulted in many pairs breeding in the twilight zone. Towards the cave mouths nests are only in shaded corners. Nests are characteristically near the top of a vertical face, within a few feet of the roof (whether in a low alcove or a tremendously high main chamber). They are often attached to a slightly overhanging face, rarely to a horizontal roof and never to a convex slope. Areas apparently suitable are unoccupied, and social as well as topographical factors must prompt the close groupings of nests. All seen at Niah were at least 2in (5cm) apart.

For many years the nests have been 'farmed'. At first the natives themselves established harvest dates – two 60-day periods from November to January, and May to July. Harvesting is now controlled by law. The first period is early in the season, and the birds rebuild; the second is at the end of the season, after fledging. The colony is in no danger; fewer nests are collected nowadays, which helps to make birds' nest soup – when it's available – one of the world's most expensive foods. Living standards at Niah have improved and fewer men are willing to climb, fewer owners ready to employ men for the job.

Nest collectors' poles in Niah Great Cave may be over 300ft tall in order to reach the ceiling. (*S. C. Bisserot FRPS*)

A collector needs a bamboo pole, a knife on the end with which to cut down the nests, ropes and climbing frames or scaffolds. He also needs to labour, sweat and be courageous. No one slips when collecting and survives, so before going up the pole is slapped for good luck. If his strength gives out, there is no help. The ceiling is 300–400ft (91–121m) up so every man risks life and limb each time he climbs. All work stops if there is an accident or a quarrel, and new sacrifices are made at the cave entrance. Collecting is a hereditary job and men have a hereditary area from which to collect on wall and roof (Attenborough 1973).

Unharvested caves are known to exist. In Java, where some birds have nested in dwelling houses, the owners have moved out, let the swiftlets take over, and have then harvested and sold the nests, and bought a new house on the proceeds.

In the Orient a couple of white nests may cost £1 or $1.70. The best restaurants vary their recipes: the nest mixed with sliced chicken, minced chicken, chicken liver or boiled chicken. It may be eaten also as the traditional soup, costing £3 or $5.10 or so for a little bowl. It is gelatinous, and strangest of all, pure birds' nest soup is – tasteless!

13 Mound builders

The breeding habits of the whole family of the *Megapodiidae* ('big-feet') are unique. The megapodes, are all ground-loving birds, confined to eastern Indonesia, Polynesia, New Guinea and Australia. There are three ecological groups: the jungle-fowl, the brush-turkeys and the mallee fowl. In times past up to 26 species were recognised, but the most recent studies suggest there are but 12 with many subspecies (Gruson 1976).

The odd breeding behaviour common to all the species is their habit of laying their eggs in a hole in the ground or in specially made mounds, making no attempt to brood them, and leaving them to be incubated by natural heat! They are collectively known as 'mound builders' or 'incubator birds'.

Several species 'nest' quite simply. The maleo fowl *Macrocephelon maleo* inhabits the forests of the northern peninsula of the island of Celebes. In the months of August and September they descend to quiet sea-beaches to lay their eggs. One particular beach will serve the scores of pairs in an extensive forest hinterland. The birds arrive in pairs at their beach, some having come 10 to 15 miles. A. R. Wallace visited one steep beach about a mile in length, of very deep, loose, coarse black volcanic sandy gravel. Above high-water mark were many holes 4–5ft (122–152cm) in diameter in which, at a depth of 1–2ft (30–60cm), the maleos' eggs were found, up to eight in each hole, several inches apart from each other, the 'clutch' being the work of several hens. These nests and those of the common scrub hen *Megapodius freycinet* and the Moluccas scrub hen *M. wallacei* are not attended by the adult birds, who return to the forest as soon as the hen has covered the eggs in sand. Apparently incubation is successfully achieved by careful selection of the nest-site. Some warmed by volcanic steam have been recorded.

The maleo fowl has weaker feet than its relations, and its claws are short and straight instead of long and greatly curved. The other mound builders are so named because with their powerful feet they scratch, dig and 'gradually manage to accumulate tumuli', wrote A. O. Hume a century ago, 'that would not have done discredit to the final resting-place of some ancient British hero'.

Cross section of a mallee fowl's mound.
c Covering of loose sand.
e Egg chamber (filled with a mixture of sandy soil and vegetation).
h Hot bed of fermenting vegetable material.

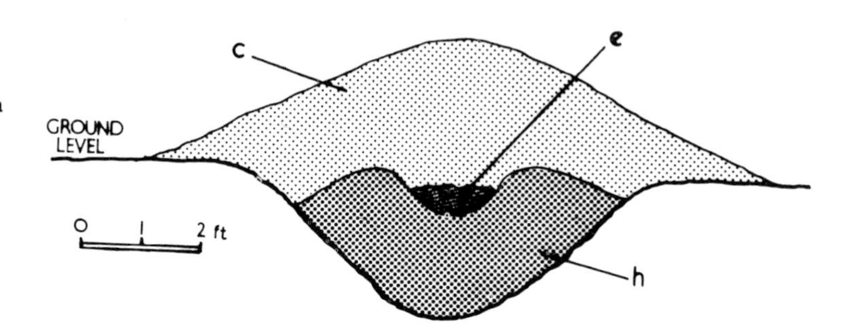

These extraordinary mounds vary in size and location according to the species. Some are dug on beaches, many are in the warm, moist jungle. In the forest, a circle of cleared ground many yards in diameter frames the mound, which is commonly 12ft (3½m) in diameter and 3ft (nearly 1m) high. Much larger ones are found and are believed to be the accumulation of many years and the work of several pairs: gigantic mounds have been measured up to 19ft (5¾m) in diameter at the base and 15ft (4½m) high. Some are a mixture of sand, soil and vegetable material; others are made purely of plant rubbish. In the latter the mound ferments rapidly and generates so much heat that with the brush turkeys *Talegalla* spp, for example, were it not for the male's activities the eggs would soon be hard-boiled. He controls the temperature by digging over the material and not until the heat has dropped is the hen allowed to lay. Each egg is laid at the end of a tunnel in the mound, as far as 3–5ft (1–1½m) below the surface. The temperature there is controlled throughout incubation, a period of up to nine weeks, by the male. The average incubation temperature in birds is 93°F (33·9°C) (Thomson 1964), which was exactly the temperature recorded in a mound of the dark-billed brush turkey *Talegalla fuscirostris* in southern New Guinea (Ogilvie-Grant 1897).

The best observed of all the mound builders is the mallee fowl *Leipoa ocellata* of arid southern Australia. For some years it was studied at the Commonwealth Scientific & Industrial Research Organisation in Griffith, New South Wales, and the fascinating story was later summarised by the research officer H. J. Frith (1962).

In the mallee fowl's dry habitat there is very little vegetable debris, and the air temperature is likely to vary considerably. Both these factors increase the problem of maintaining a satisfactory constant temperature. The birds' answer to the problem is a more complicated nesting behaviour than any of the other megapodes. Firstly a hole is dug in the ground up to 15ft (4½m) in diameter and a few feet deep. This is filled in winter with vegetable material, swept in from a radius of 50yd (45m) or so. After being soaked in spring rain this is covered thickly with a mound of sand, under which it ferments. In a year with little or no rain the birds will not nest; but if all is well the male digs a hollow in the rotting plant material, into which all the eggs are laid, at intervals of about a week, up to two dozen in all. In several nests white ants have been found, making little covered galleries around and attached to the eggshells.

The male uncovers and covers the mound for each laying and constantly controls the mound's temperature. When fermentation is considerable in spring the mound is opened to allow heat to escape. Later the vegetation's heat-producing ability is much reduced, but the sun then warms the mound so much that it has to be built up to insulate the eggs from the outside heat. If the temperature is still too high the soil may be scraped off early in the morning, spread around to cool, and then replaced; in colder weather it is spread at midday to warm-up and replaced later in the afternoon. The male repeatedly pecks and digs into its mound and by so doing is believed to be able to check the temperature by the use of its tongue or the sensitive

A pair of mallee fowl *Leipoa ocellata*, male left, at their mound in Australia. These birds excavate a large hole in the ground which they fill with vegetable matter. After the spring rains fermentation begins and the eggs, up to a dozen, are laid. These eggs are not incubated by the parent birds but by the natural heat of the mound and its fermentation. Almost immediately after hatching the young are able to look after themselves and they are never cared for or even recognised by the parent birds. (*Australian News and Information Bureau*)

inside of its bill. Experiments by Dr Frith proved that the male could regulate his mound's temperature for eight months of the year to within one degree of 92°F (33·3°C) – an incredible feat of engineering.

Unlike many of its relatives the mallee fowl is a very sedentary bird, some males spending all their lives within a few hundred yards of their mounds.

Mound builders lay very large eggs. A maleo is only 22in (56cm) long and her egg measures $4\frac{1}{4} \times 2\frac{1}{2}$in (about 11×6cm); a mallee fowl is 24in (61cm) long and lays an egg $3\frac{1}{2} \times 2\frac{1}{4}$in ($8\frac{3}{4} \times 5\frac{3}{4}$cm). The latter's egg represents about one-tenth of the hen's body weight.

Most megapodes' eggs are laid with the narrow end pointing downwards, and those of several species have been described as being thin-shelled, which is curious considering the way they are buried. Many native tribes regard the birds and their eggs as a delicacy, even to the extent of carefully farming the eggs. Scrub-fowl eggs have been described as 'equal, if not superior, to those of the Pea Fowl . . . and higher commendation cannot be given', and the maleo's eggs 'are delicious eating, as delicate as a fowl's egg, but much richer' (Ogilvie-Grant 1897).

From such large eggs emerge fully feathered young. Each egg begins to incubate as soon as it is laid, so hatching may take place throughout a period of several months. The chicks dig their own way out of the mounds, are usually dry when they break surface, can move about straight away, run swiftly and feed within a few hours, and are able to fly within 12–24 hours. As with all other mound builders the young and adults do not appear to recognise each other, and the cock and hen certainly make no attempt to care for their chicks.

Despite all these complicated and energy-consuming activities – a male mallee fowl is building or maintaining his mound for 11 months – the birds' breeding success is not as high as one might expect. Predators are common and of 1,094 mallee-fowl eggs noted by Dr Frith only 542 hatched successfully ($49\frac{1}{2}$ per cent) whereas the fledging success of cavity-nesting birds is on average 66 per cent.

The life histories of the other megapodes are not well known. Just why the family has evolved such a unique nest is a puzzle which deserves unravelling. One suggestion is that there is insufficient food to support adults and fledged young if all are foraging together, but by staggering the hatching of young the food supply is shared more efficiently.

The first western explorer to see an incubator bird was a member of Magellan's expedition round the world in 1521. Since then the more we have come to learn of the birds the more incredible they seem. Cayley (1950) remarked that the mallee fowl was threatened by farming; 20 years later it still had to be stated that only 'the provision of adequate reserves will ensure its survival' (Serventy 1969). The jungle-loving species from further north are safe so long as the rain-forests continue to remain untouched. But at the time of writing (1976) the World Wildlife Fund is engaged in Operation Rainforest – an attempt at money-raising to help preserve the world's rapidly diminishing tropical forests. It is to be hoped that these fascinating birds will long survive to intrigue everyone interested in wildlife.

14 Parasites

Having looked at plovers which have no nests yet care for their young, and mound-builders which build huge nests but take no part in rearing their young, we come to birds which neither build a nest nor rear their chicks, yet breed successfully – the parasites.

A parasite is an organism which lives at the expense of another, commonly in or upon it; its host receives no advantages from the relationship, and indeed often suffers because of it. Birds have many parasites living on them, such as ticks, feather-mites, louse-flies, fleas and lice. A rook *Corvus frugilegus* may be host to six species of lice alone, quite apart from the nematode worms and flukes which may be living *in* its body. But no bird is itself wholly parasitic. Some do appropriate the nests of other species to rear their own brood, and there are birds – such as the skuas or jaegers – which rob others of their food (or young), but this is more usually referred to as mere 'piracy'. However, a few birds do show real parasitism in that their eggs are laid in the nests of other species, and the young brought up by the unwitting hosts. This is known as nest parasitism, the prize example of which is the breeding style of the cuckoo *Cuculus canorus*, the bird which gave its name to the whole family and gave English the word 'cuckold', although curiously the word is applied to the 'host' rather than the villain of the piece. In the animal kingdom the only other nest-parasites are some insects (known, obviously, as cuckoo-bees and cuckoo-flies) which foist their family on other insects.

Any bird which is a nest-parasite must have a life-style which matches its host's. Their food requirements must be similar; the host must be a common species, its nest easy to find and enter; their incubation and fledging periods must be about the same; and the parasite must be able to time its egg-laying to coincide with that of the host. Otherwise the parasite will not breed successfully. Certain species have evolved special characteristics which improve their chances of survival: cuckoos lay eggs which mimic those of the host; the young instinctively reject the chicks or eggs of the host; and some young parasites mimic the appearance of their foster-brothers.

Fischer's whydah *Vidua fischeri* is host-specific, choosing to lay its eggs only in the nests of the purple grenadier *Estrilda ianthinogaster*. The young whydah matures side by side with its foster brothers and competes with them for food on equal terms, only surviving because of its detailed mimicry.
(J. F. Reynolds)

The stifftails are a rather strange tribe of ducks. Of the nine species, eight build elaborate nests in the reeds, but the ninth, the black-headed duck *Heteronetta atricapilla* of parts of Chile, Paraguay, southern Brazil and Argentina, is a nest-parasite. Little is known of its life-history, but its eggs have been found in the nests of other ducks, coots and rails, herons and even a carrion-eating bird of prey, one of the cacacaras *Polyborus* spp. This parasitism may be a recent and almost accidental change of habit (Southern 1964), and it is difficult to see that it could be successful enough when a host so radically different from the duck has been chosen.

Honeyguides are rather dull grey-brown birds, nearly as big as starlings, which inhabit southern Asia and Africa south of the Sahara. They guide mammals, including man, to bees' nests and when the honey is collected the honeyguide eats the beeswax. Of the 14 species, the breeding habits of one, the lyre-tailed *Melichneutes robustus*, are unknown (the Congo forest natives call it 'the bird nobody ever sees') but the rest are all parasitic. For example, the lesser honeyguide *Indicator minor* lays in the tree-hole nest of its close relative, a wood-

This fork-tailed drongo *Dicrurus adsimilis*, seen with its hammock-shaped nest and chick, is quite likely to play unwitting host to the young of a drongo cuckoo *Surniculus lugubris*. (*J. B. & S. Bottomley*)

pecker, or more often a barbet. The eggs are white, as are those of most hole-nesting species, although how many each bird lays is not known.

Instinctively eliminating competition for food, the young parasite usually kills its nest mates and, if it can, will push them out of the nest. To assist it in the murder, the assassin's bill has a needle-sharp hook at the tip of each mandible, which drops off after about ten days. The hosts then rear the chick as their own. Yet when the honeyguides are prospecting, the barbets defend their nest zealously: honeyguides return again and again, the male trying to lure the rightful owners away from the hole so that, in an unguarded moment, the hen can slip in and leave an egg. Struggles have been known to last all day (Friedmann 1955).

The only parasites in North America are the cowbirds *Molothrus ater*. Outside the breeding season these are gregarious birds, moving south in flocks in winter, pairs taking up territories in the north again in spring. Many species, especially sparrows, warblers and vireos are their victims. The cowbirds' eggs have an incubation period of only 10 to 12 days, whereas the host birds' eggs average a few days longer, so the young cowbird often hatches first or is not far behind its nest-mates. It does not reject the other young, but some of them suffer simply because they are smaller and weaker. A few host species throw out a cowbird's egg; some desert the nest, disturbed by the parasite and probably the strange egg (for cowbirds do not show egg mimicry); and the yellow warbler *Dendroica petechia* ingeniously floors over the cowbird's egg before laying her own; as many as four cowbird eggs have been successively covered. But once a host has a cowbird chick it feeds it as solicitously as it would its own, even for some time after fledging.

The cowbird's breeding success is believed to be rather low, but each hen lays about 25 eggs which adequately maintain the species. The bird is probably increasing because it is evolving the habit of parasitising more species, including basically forest birds; its own habitat used to be only grassland. The populations of the hosts are not adversely affected because they are usually well able to rear several of their own young.

In South America the shiny cowbird *M. bonariensis* also victimises many small passerines, even tyrant flycatchers which are very aggressive towards all other birds near their nests. It is believed to lay the astonishing number of 60–100 eggs per season. The giant cowbird *Scaphidura oryzivora*, also found in much of South America's open country, is very selective, laying in the pendent nests of caciques and oropendolas, so that a close study of the birds' breeding habits is almost impossible.

A female cuckoo *Cucullus canorus* (known as a reed warbler cuckoo because of its host preference) flies towards a reed warbler's nest on a reconnaissance trip. (*I. Wyllie*)

A female cuckoo photographed while removing an egg from a reed warbler's nest. (*I. Wyllie*)

Finally, the screaming cowbird *M. rufoaxillaris* is believed to choose only the bay-winged cowbird *M. badius* as a host, and that species is not parasitic at all! So the student of the evolution of nest parasitism in birds has the whole range in one family.

In the big weaver-bird family is one small yellow species, the cuckoo weaver *Anomalospiza imberbis*, which is parasitic. Apart from the fact that its hosts are grass-warblers *Cisticola* spp, nothing is known of its life.

Little more is known of the parasitic African weavers of the genus *Vidua*. This is surprising because several of them are quite common. The relationships of the various species to each other are not clear, but studies so far have shown that both the indigo birds and the whydahs are parasites only of the little weaver-finches *Estrilda* spp. What is more, those that are better known victimise particular species: that is, they are host-specific. For example, the paradise whydah *V. paradisaea* victimises the melba finch *Pytilia melba*, the shaft-tailed whydah *V. regia* chooses the violet-eared waxbill *Estrilda granatina*, and Fischer's whydah *V. fischeri* only lays in the nests of the

purple grenadier *E. ianthinogaster*. The last two are particularly interesting because the ranges of the parasites coincide almost exactly with those of their hosts. Young whydahs grow up side by side with their foster brothers and sisters, and compete on equal terms for food. To assist them each species has mimicked in extraordinary detail the palate and gape markings of the weaver-finches – the bright special markings that provide the stimulus for the parents to feed the young birds. Without them the whydahs would not survive.

There are nearly 130 cuckoos, but only 47 in the subfamily which contains the parasitic species, all of which are confined to the Old World. Indeed, just to add interest or confusion, some cuckoos with the name 'cuckoo' are not parasitic, such as the American black-billed and yellow-billed cuckoos *Coccyzus erythrophthalmus* and *C. americanus*; and some cuckoos not called 'cuckoo' *are* parasitic, like the koel *Eudynamys scolopacea* of Asia. The parasitic cuckoos range widely in size, and this is reflected somewhat in their host species: for example the little bronze cuckoo *Lamprococcyx minutillus* of Australia is only just over 6in (15cm) long and parasitises warblers (chiefly *Gerygone* spp) which are even smaller; whereas the koel (16½in, 42cm) and the handsome great spotted cuckoo *Clamator glandarius* (15½in, 39cm) victimise magpies and crows.

Some kinds of young cuckoos grow up with the young of the foster parents, others eject them. In the former group is the drongo cuckoo *Surniculus lugubris*, of south-east Asia, so called because not only do the adults mimic the black plumage of the drongos, *Dicrurus* spp, but remarkably the immature plumages are similar too; both species have dark feathers tipped with white. Drongos are well known for the noisy, energetic way in which they will usually defend their nests, so the young cuckoos' plumage is presumably an adaptation to make them acceptable to the drongos. This arrangement looks less tidy and obvious when it is realised that drongo cuckoos also choose hosts which are not black. The second group is best exemplified by the cuckoo *Cuculus canorus*, which is a widespread breeder in the Palaearctic, migrating south for the winter.

Few birds have a documented history as long as the cuckoo's. Its two-note call which heralds the spring and its strange nesting behaviour have intrigued men for ages. The first written record we have is from Aristotle (384–322 BC): 'The cuckoo lays in a nest which she has not built herself, but of some smaller bird, eating the eggs she finds there, and leaving her own.' A more complete account came from the Roman army officer, Pliny the Elder (AD 23–79), who recorded that the cuckoo often chose a particular host, laid one egg per nest, and left the host to rear the young one. He quaintly added that 'The reason why they have other birds to sit upon their eggs and hatch them is because they know how all birds hate them. For fear, therefore, that the whole race be utterly destroyed by the fury of others of the same kind, they make no nest of their own and are forced by this crafty shift to avoid danger.'

For long the species' life continued to be known by a mixture of fact and legend. In England, its similar appearance to the sparrowhawk *Accipiter nisus* led to the belief that the hawks turned into cuckoos in the summer, and for reasons which are not obvious the cuckoo was widely believed to be a foolish bird; it was, and still is, the custom to call a foolish person a cuckoo. Finally 'The Cuckoo Song' is at least seven centuries

A reed warbler's nest showing the warbler's clutch of three eggs and one, slightly larger, cuckoo's egg. (*I. Wyllie*)

Right:
A reed warbler struggles feed its foster chick, a 16-day-old cuckoo already larger than its adopted parent. Warblers rear their own young for from 10–14 days but will continue to feed a young cuckoo for twice that time. (*I. Wyllie*)

Below:
A scrawny, hairless one-day-old cuckoo is already old enough and strong enough to eject the competition – in this case a reed warbler's egg. (*I. Wyllie*)

old but is still sung in England today; the oldest extant version begins:

> Sumer is icumen in
> Lhude sing cuccu!
> Groweth sed and bloweth med
> And springth the wde nu.
> Sing cuccu!

Even in the nineteenth century when more real information about the cuckoo was discovered much was still misunderstood. Saunders (1899) in his great work confidently wrote that a hen cuckoo laid her egg on the ground and then conveyed it to the nest of the foster-parent, and it was not until Chance (1922) published his detailed observations that the truth was understood, and found to be just as intriguing as the myths. The female maintains a territory and tries to deter other females from using the area. Each female is usually host-specific and sometimes cuckoos are therefore known as reed warbler-cuckoos, meadow pipit-cuckoos or dunnock cuckoos, those three hosts being the birds most commonly victimised in Britain. Over 50 hosts have been recorded including the tiny wren which builds an enclosed nest that must present quite a problem to a cuckoo ready to lay.

Once the hen has located the nests to be used she lays an egg directly into them, one every two days, spending only about nine seconds at the nest to deposit the egg (Seel 1973); this unusually short time is presumably an adaptation to disturb the hosts as little as possible in order to ensure they do not desert. She takes no further interest in the chosen nest, each subsequent egg being laid elsewhere. So the number of eggs laid is dependent upon the number of nests the cuckoo can find. Chance proved (under rather artificial conditions) that meadow pipit-cuckoos laid up to 25 eggs in a season, but the largest number of eggs laid by a reed warbler-cuckoo in one particular study season was 12 (Wyllie 1975). Usually the hen cuckoo removes just one egg from the host nest.

Each cuckoo's egg differs in size, colour and markings from those of other cuckoos; these differences are so characteristic that those of each female can be identified. A very high degree of mimicry is also to be found in the eggs, birds which are particularly host-specific laying eggs which look uncannily like the host's, although glaring differences may also be found: for example the dunnock *Accentor modularis* lays plain sky-blue eggs but the cuckoo's egg is often speckled. A cuckoo usually lays her egg after midday, unlike most birds which lay early in the morning. This adaptation enables her to reconnoitre the area thoroughly and to add her egg when the host-species is actually forming its clutch, preferably when only two eggs have been laid; the cuckoo's egg will then hatch (after 12 days) a little before, or only just after, those of its nest mates.

The young cuckoo ejects its host's eggs or young between 8 and 36 hours after hatching, and is then fed for about three weeks by its foster-parents, and for some days after fledging, when its loud insistent begging-call even attracts other birds to feed it! It is curious that a reed warbler, for example, rears its own young in 10 to 14 days, but will carry on feeding a cuckoo for twice as long. Little is known of the cuckoo's breeding success, but after passing up the responsibility of parental care all is not plain sailing, as Wyllie (1976) discovered: during one season in a reed bed in Cambridgeshire, 136 reed warblers' nests were found and 25 of these were cuckolded, yet only one fledged young was seen – although an unlocated nest produced a fledged cuckoo as well.

We have learned a great deal about the cuckoo, but many questions remain. Many books state that the male is polygamous, but recent studies suggest that polyandry or promiscuity are practised. How helpful *is* egg-mimicry? How successful are cuckoos compared with cowbirds? Why aren't New World cuckoos parasitic? The more we know, the more we realise that we do not understand.

Appendix: Additional species of interest

The following is a list of other birds whose nesting is worth reading about, and details of which would fill another book!

Chapter 2 No nest
Laysan albatross, the Midway Island 'gooney bird' *Diomedea immutabilis*
great bustard *Otis tarda*
stone curlew *Burhinus oedicnemus*
dotterel *Eudromias morinellus*
emu *Dromaius novaehollandiae*
fulmar *Fulmarus glacialis*
loon *Gavia immer*
snowy owl *Nyctea scandiaca*
partridge *Perdix perdix*
ptarmigan *Lagopus mutus*

Chapter 3 Simple constructions
whooping crane *Grus americana*
yellow-billed cuckoo *Coccyzus americanus*
collared dove *Streptopelia decaocto*
bald eagle *Haliaeëtus leucocephalus*
Gould's frogmouth *Batrochostoma stellatus*
gannets, boobies *Sula* spp.
lammergeier *Gypaetus barbatus*
marbled murrelet *Brachyramphus marmoratus*
great horned owl *Bubo virginianus*
mute swan *Cygnus olor*

Chapter 4 Cupshaped nests
bowerbirds *Ptilonorhynchidae*
bullfinch *Pyrrhula pyrrhula*
canary *Serinus canaria*
fantails *Rhipidura* spp.
helmet-shrikes *Prionops* spp.
red-billed magpie *Urocissa erythrorhyncha*
robin *Erythacus rubecula*
Australian robins *Petroica* spp.
shrike-tit *Falcunculus frontatus*

Chapter 5 Enclosed nests
Australian babblers *Pomatostomus* spp.
Spinetails *Synallaxis* spp.
yellow-tailed thornbill *Acanthiza chrysorrhoa*
fairy warbler *Gerygone flavida*
black-rumped waxbill *Estrilda troglodytes*

blue wrens *Malurus* spp.
bananaquit *Coereba flaveola*

Chapter 6 Woven and hanging nests
African broadbill *Smithornis capensis*
crombecs *Sylvietta* spp.
eremomelas *Eremomela* spp.
royal flycatcher *Onychorhynchus coronatus*
mistletoe bird *Dicaeum hirundinaceum*
sunbirds *Nectarinia* spp.
red-eyed vireo *Vireo olivaceus*
many white-eyes *Zosteropidae*

Chapter 7 Mud nests
black-browed albatross *Diomedea melanophris*
cock-of-the-rock *Rupicola peruviana*
torrent-lark *Grallina bruijni*
many swallows *Hirundinidae*

Chapter 8 Ground nests
bluethroat *Luscinia svecica*
Lapland bunting *Calcarius lapponicus*
coucal *Centropus* spp.
skylark *Alauda arvensis*
Lincoln's sparrow *Zonotrichia lincolni*
whinchat *Saxicola rubetra*

Chapter 9 In holes
budgerigar *Melopsittacus undulatus*
hoopoe *Upupa epops*
kiwi *Apteryx australis*
red-breasted nuthatch *Sitta canadensis*
burrowing owl *Athene cunicularia*
swallow tanager *Tersina viridis*
tapaculos *Rhinocryptidae*
inca tern *Larosterna inca*
toco toucan *Ramphastos toco*

Chapter 10 Aquatic nests
bittern *Botaurus stellaris*
giant pied-billed grebe *Podilymbus gigas*
western grebe *Aechmophorus occidentalis*

Chapter 11 Colonial nesting
black-browed albatross *Diomedea melanophris*
Eleanora's falcon *Falco eleonorae*
frigate-birds *Fregata* spp.
swallow-tailed gull *Creagrus furcatus*
heronries
kittiwake *Rissa tridactyla*
bald ibis *Geronticus calvus*
scarlet ibis *Eudocimus ruber*
pelicans *Pelecanidae*
Galapagos penguin *Spheniscus mendiculus*
rook *Corvus frugilegus*

Bibliography

Alder, James (1963) 'Behaviour of dippers at the nest during a flood', *British Birds* 56: 73–6.
Andrade, C. S. (1969) in *Birds of the World*, ed. J. Gooders, p. 1673.
Armstrong, Edward A. (1947) *Bird Display and Behaviour*. London.
Armstrong, Edward A. (1955) *The Wren*. London.
Ashmole, N. P. (1962) 'The black noddy on Ascension Island'. *Ibis*, 103b: 236–8.
Ashmole, N. P. (1963) 'The biology of the wideawake or sooty tern', *Ibis*, 103b: 298–364.
Attenborough, David (1973) 'Zoo Quest in Borneo'. TV film for British Broadcasting Corporation.
Austin, Oliver L. (1961) *Birds of the World*. London.
Baker, E. C. S. (1942) *Cuckoo Problems*. London.
Bannerman, D. A. (1953) *The Birds of the British Isles*, 1. Edinburgh.
Bannerman, D. A. (1956) *The Birds of the British Isles*, 5. Edinburgh.
Bannerman, D. A. (1957) *The Birds of the British Isles*, 6. Edinburgh.
Bird, G. (1933) 'Breeding biology of the little grebe', *British Birds*, 27: 34–7.
Blair, H. (1962) in *The Birds of the British Isles* by D. A. Bannerman, 11. Edinburgh.
Boswall, J. (1976) 'Safari to Mexico', TV film for British Broadcasting Corporation.
Brosset, A. (1969) in *Birds of the World*, ed. J. Gooders.
Brown, Leslie (1970) *Eagles*. London.
Brown, Leslie and Amadon, Dean. (1968) *Eagles, Hawks and Falcons of the World*, 2. Feltham.
Cayley, Neville W. (1950) *What Bird Is That?* Sydney.
Chance, E. (1922) *The Cuckoo's Secret*. London.
Coward, T. A. (1950) *The Birds of the British Isles and their Eggs*, vols. 1 and 2, seventh edition. London.
Crook, John H. (1960) 'Nest form and construction in certain West African weaver birds', *Ibis*, 102: 1–25.
Crook, John H. (1963) 'A comparative analysis of nest structure in the weaver birds', *Ibis*, 105: 238–62.
Cullen, J. M. and Ashmole, N. P. (1963) 'The black noddy on Ascension, part 2', *Ibis*, 103b: 423–9.
Darling, F. Fraser (1952) *Bird Flocks and the Breeding Cycle*. Cambridge.
Davies, S. J. J. F. (1962) 'The nest-building behaviour of the magpie goose', *Ibis*, 104: 147–57
Dorward, D. F. (1963) 'The fairy tern at Ascension Island', *Ibis*, 103b: 365–78.
D'Urban, W. S. M. and Mathew, Murry A. (1895) *The Birds of Devon*, second edition. London.

Durrell, Gerald (1956) *The Drunken Forest*. London.
Fisher, James (1952) *The Fulmar*. London.
Freuchen, Peter and Salomonsen Finn, (1959) *The Arctic Year*. London.
Friedmann, H. (1949) 'The breeding habits of the weaver birds', *Annual Report Smithsonian Institution*, 1949, pp. 293–316.
Friedmann, H. (1955) *The honey-guides*. U.S. National Museum Bulletin, 208.
Frith, H. J. (1962) *The Mallee Fowl*. Sydney.
Gaston, A. J. (1973) 'The ecology and behaviour of the long-tailed tit', *Ibis*, 115: 330–51.
Gill, E. Leonard (1936) *A First Guide to South African Birds*. Cape Town.
Gooders, John (1969–71) Editor, *Birds of the World*, encyclopedia.
Goodfellow, Peter (1973) *Projects with Birds*. Newton Abbot.
Gordon, Seton (1955) *The Golden Eagle*. London.
Greenway, James C. (1967) *Extinct and Vanishing Birds of the World*, second edition. New York.
Grimes, L. G. (1976) 'The occurrence of co-operative breeding behaviour in African birds', *The Ostrich*, 47: 1–15.
Gruson, Edward S. (1976) *Checklist of the Birds of the World*. London.
Hall, K. R. L. (1960) 'Egg covering by the white-fronted plover', *Ibis*, 102: 545–53.
Harris, Michael (1974) *A Field Guide to the Birds of Galapagos*. London.
Harrison, Colin (1975) *A Field Guide to Nests, Eggs and Nestlings of European Birds*. London.
Harrison, C. J. O. (1967) 'Sideways throwing and sideways building in birds', *Ibis*, 109: 539–51.
Harrison, C. J. O. (1973) 'Nest building behaviour of quaker parrots', *Ibis*, 115: 124–8.
Harrison, Jeffery (1973) *A Wealth of Wildfowl*, revised edition. London.
Herrick, Francis H. (1911) 'Nests and nest-building in birds', *Journal of Animal Behaviour*, 1: 159–92, 244–77, 336–73.
Howes, Colin A. (1969) 'Breeding behaviour of the blue tit', *Devon Birds*, 22: 3–11.
Hudson, Robert (1975) *Threatened Birds of Europe*. London.
Jameson, William (1958) *The Wandering Albatross*. London.
Johnson, A. W. (1969) in *Birds of the World*, ed. J. Gooders, p. 790.
Jones, Peter Hope and Dare, Peter (1976) *The Birds of Caernarvonshire*. Llandudno.
Kahl, M. P. (1967) 'Observations on the behaviour of the hammerkop', *Ibis*, 109: 25–32.
Lack, David (1956) *Swifts in a Tower*. London.
Lack, David (1966) *Population Studies of Birds*. Oxford.
Levick, G. M. (1914) *Antarctic Penguins*. London.
Mackworth-Praed, C. W. and Grant, C. H. B. (1973) *Birds of West Central and Western Africa*. London.
Maclaren, P. I. R. (1950) 'Bird-ant nesting associations', *Ibis*, 92: 564–6.
Maclean, G. L. (1967) 'The breeding biology of the double-banded courser', *Ibis*, 109: 556–69.

Madge, S. G. (1970) 'Nest of the long-billed spider hunter', *Malayan Naturalists' Journal*, 23: 125.
Makatsch, W. (1950) *Der Vogel und Sein Nest*. Leipzig.
Medway, Lord (1962) 'The swiftlets of Niah Cave, Sarawak', *Ibis*, 104: 45–66.
Moreau, R. E. (1941a) 'A contribution to the breeding biology of the palm swift', *Journal of the East Africa & Uganda Nat. Hist. Soc.*, 15: 154–70.
Moreau, R. E. and W. M. (1941b) 'Breeding biology of the silvery-cheeked hornbill', *Auk* 58: 13–27.
Moreau, R. E. (1960) 'Conspectus and classification of the Ploceine weaver-birds (concluded)', *Ibis*, 102: 443–71.
Murphy, Robert C. (1959) 'Peru profits from sea fowl', *National Geographic Magazine*, 115: 395–413.
Nelson, Bryan (1968) *Galapagos, Islands of Birds*. London.
Nethersole-Thompson, Desmond (1975) *Pine Crossbills*. Berkhamsted.
Ogilvie, M. A. (1975) *Ducks of Britain and Europe*. Berkhamsted.
Ogilvie-Grant, W. R. (1897) *A Handbook to the Game-birds*, 2. London.
Owen, J. H. (1945) 'The Long-tailed tit', *British Birds*, 38.
Peterson, Roger Tory (1961) *A Field Guide to Western Birds*. Boston.
Phillips, W. W. A. (1952) *Birds of Ceylon*, 2. Colombo.
Phillips, W. W. A. (1961) *Birds of Ceylon*, 4. Colombo.
Pollard, A. W. (1900) Editor, *The Travels of Sir John Mandeville*. London.
Rankin, Niall (1947) *Haunts of British Divers*. London.
Rennie, J. (1831) *The Architecture of Birds*. London.
Richardson, F. (1965) 'Breeding and feeding habits of the black wheater of southern Spain', *Ibis*, 107: 1–16.
Roberts, Austin (1940) *The Birds of South Africa*. Cape Town.
Root, Joan and Alan (1969) 'Inside a hornbill's walled-up nest', *National Geographic Magazine*, 136: 846–55.
Ross, Edward S. (1965) 'Birds that "see" in the dark with their ears', *National Geographic Magazine*, 127: 282–90.
Ryves, B. H. (1944) 'Nest-construction by birds', *British Birds*, 37: 182–8, 207–9.
Sage, B. L. (1969) 'Breeding biology of the coot', *British Birds*, 62: 134–43.
Saunders, Howard (1899) *Manual of British Birds*. London.
Scheithauer, Walter (1967) *Hummingbirds*. London.
Seel, D. C. (1973) 'Egg-laying by the cuckoo', *British Birds*, 66: 528–35.
Serventy, Vincent (1969) in *Birds of the World*, ed. J. Gooders, p. 625.
Sharpe, R. Bowdler (1898) *Wonders of the Bird World*. London.
Sharrock, J. T. R. (1976) Editor, *The Atlas of Breeding Birds in Britain and Ireland*. Tring.
Sielmann, Heinz (1959) *My Year with the Woodpeckers*. London.
Simmons, K. E. L. (1955) 'Studies on great crested grebes', *The Avicultural Magazine*, 61: 235–53.
Simmons, K. E. L. (1972) 'Some adaptive features of seabird types', *British Birds*, 65: 465–79, 510–21.

Skead, C. J. and Ranger, G. A. (1958) 'A contribution to the biology of the Cape Province white-eyes', *Ibis*, 100: 327.
Skutch, Alexander F. (1940) 'Social and sleeping habits of Central American wrens', *Auk*, 57: 293–312.
Skutch, Alexander F. (1958) 'Life history of the white-winged soft-wing', *Ibis*, 100: 213–24.
Skutch, Alexander F. (1962) 'Life history of the white-tailed trogon', *Ibis*, 104: 301–13.
Skutch, Alexander F. (1963) 'Life history of the rufous-tailed jacamar in Costa Rica', *Ibis*, 105: 354–68.
Skutch, Alexander F. (1964) 'Life history of the blue-diademed motmot', *Ibis*, 106: 321–32.
Skutch, Alexander F. (1973) *The Life of the Hummingbird*. London.
Smith, Stuart (1950) *The Yellow Wagtail*. London.
Snow, D. W. (1958) *A Study of Blackbirds*. London.
Southern, H. N. (1964) Article 'Parasitism' in Thomson, A. L. (ed.), *New Dictionary of Birds*. London & New York.
Sparks, John and Soper, Tony (1967) *Penguins*. Newton Abbot.
Summers-Smith, D. (1963) *The House Sparrow*. London.
Thomson, A. Landsborough (1964) *A New Dictionary of Birds*. London.
Tinbergen, N. (1953a) *Social Behaviour in Animals*. London.
Tinbergen, N. (1953b) *The Herring Gull's World*. London.
Truslow, F. R. (1960) 'Return of the trumpeter', *National Geographic Magazine*, 118: 134–50.
Tweedie, M. W. F. (1960) *Common Malayan Birds*. London.
Van Tyne, Josselyn and Berger, Andrew J. (1959) *Fundamentals of Ornithology*.
Vesey-Fitzgerald, D. F. (1958) 'Notes on breeding colonies of the red-billed quelea in S.W. Tanganyika', *Ibis*, 100: 167–9.
Voous, K. H. (1960) *Atlas of European Birds*. London.
Wetmore, Alexander (1934) 'Winged denizens of woodland, stream and marsh', *National Geographic Magazine*, 65: 577–96.
Wood, J. G. (1892) *Wonderful Nests*. London.
Wyllie, Ian (1975) 'Study of cuckoos and reed warblers', *British Birds*, 68: 369–78.

Index